Lecture Notes in Mathematics Vol. 989
ISBN 978-3-540-12294-4 © Springer-Verlag Berlin Heidelberg 2008

Angelo B. Mingarelli

Volterra-Stieltjes Integral Equations and Generalized Ordinary Differential Expressions

Errata

- p.90: Proof of Theorem 2.3.1; delete the remark in parentheses "since -$y(t)$ is a solution" (comment: clearly this is not assumed; if $y(t) < 0$ in (2.3.11) define $g(t)$ by the negative of the right-side of (2.3.12))
- p.244, Statement of Theorem 5.2.2: Replace the word "precisely" by "contained in" (comment: as this is what is proved there).

Lecture Notes in Mathematics

Edited by A. Dold and B. Eckmann

989

Angelo B. Mingarelli

Volterra-Stieltjes Integral Equations and Generalized Ordinary Differential Expressions

Springer-Verlag
Berlin Heidelberg New York Tokyo 1983

Author

Angelo B. Mingarelli
Department of Mathematics, University of Ottawa
585 King-Edward Avenue, Ottawa, Ontario, Canada K1N 9B4

AMS Subject Classifications (1980): Primary: 45 J 05, 45 D 05, 47 A 99
Secondary: 34 B 25, 34 C 10, 39 A 10, 39 A 12, 47 B 50

ISBN 3-540-12294-X Springer-Verlag Berlin Heidelberg New York Tokyo
ISBN 0-387-12294-X Springer-Verlag New York Heidelberg Berlin Tokyo

Printing and binding: Beltz Offsetdruck, Hemsbach/Bergstr.
2146/3140-543210

Quest' opera è umilmente dedicata
ai miei cari genitori Giosafat e
Oliviana e al mio fratello Marco

A.M.D.G.

PREFACE

The aim of these notes is to pursue a line of research adopted by many authors (W. Feller, M.G. Krein, I.S. Kac, F.V. Atkinson, W.T. Reid, among others) in order to develop a qualitative and spectral theory of Volterra-Stieltjes integral equations with specific applications to real ordinary differential and difference equations of the second order.

We begin by an extension of the classical results of Sturm (comparison theorem, separation theorem) to this more general setting. In chapter 2 we study the oscillation theory of such equations and, in Chapters 3,4,5, apply some aspects of it to the study of the spectrum of the operators generated by certain generalized ordinary differential expressions associated with the above-mentioned integral equations.

In order to make these notes self-contained some appendices have been added which include results fundamental to the main text. Care has been taken to give due credit to those researchers who have contributed to the development of the theory presented herein - any omissions or errors are the author's sole responsibility.

I am greatly indebted to Professor F.V. Atkinson at whose hands I learned the subject and I also take this opportunity to acknowledge with thanks the assistance of the Natural Sciences and Engineering Research Council of Canada for continued financial support. My sincere thanks go to Mrs. Frances Mitchell for her expert typing of the manuscript.

Finally, I am deeply grateful to my wife Leslie Jean for her constant encouragement and patience and I also wish to thank Professor A. Dold for the possibility to publish the manuscript in the Lecture Note series.

Angelo B. Mingarelli
Ottawa, April 1980.

TABLE OF CONTENTS

Page

CHAPTER 3 (continued)

INTRODUCTION

Let $p,q: I \to \mathbb{R}$, $p(t) > 0$ a.e. (in the sense of Lebesgue measure) and $1/p,\ q \in L(I)$ where $I = [a,b] \subset \mathbb{R}$. Consider the formally symmetric differential equation

$$(p(t)y')' - q(t)y = 0, \qquad t \in I. \tag{1}$$

By a solution of (1) we will mean a function $y: I \to \mathbb{C}$, $y \in AC(I)$, (i.e., absolutely continuous on I) such that $py' \in AC(I)$ and $y(t)$ satisfies (1) a.e. on I. Let $\gamma \in I$. Then a quadrature gives, for $t \in I$,

$$p(t)y'(t) = \beta + \int_{\gamma}^{t} y(s)q(s)ds$$

where $\beta = (py')(\gamma)$. Since $q \in L(I)$ its indefinite integral $\sigma(t) = \int_a^t q(s)ds$ exists for $t \in I$ and $\sigma \in AC(I)$. Hence y will be a solution of (1) if and only if $y(t)$ satisfies a Stieltjes integro-differential equation of the form

$$p(t)y'(t) = \beta + \int_{\gamma}^{t} y(s)d\sigma(s), \qquad t \in I, \tag{2}$$

where the integral may be interpreted, say, in the Riemann-Stieltjes sense. On the other hand (2) also has a meaning whenever $\sigma \in BV(I)$ (i.e., bounded variation on I) and y is continuous on I. Hence equations of the form (2) may be used to deal with differential equations (1). Moreover σ need not be continuous on I (as long as we require a solution of (2) to be continuous on I) and so (2) can be used to treat discrete problems, e.g., difference equations (or three-term recurrence relations) as well as continuous problems as we have seen. More precisely let $t_{-1} = a < t_0 < t_1 \ldots < t_{m-1} < t_m = b$ be a fixed partition of I. Define step-functions $p, \sigma \in BV(I)$ as follows: p, σ will be right-continuous on I and their only jumps, if any, will be at the points $\{t_i\}$ defined above with the saltus of σ being given by

$$\sigma(t_n) - \sigma(t_n - 0) = b_n,$$

where b_n, $n = 0,1,\ldots,m-1$, is a given real finite sequence, and p is defined on $[a,b]$ upon setting

$$p(t) = C_{n-1}(t_n - t_{n-1}), \qquad t \in [t_{n-1}, t_n)$$

where C_{n-1}, $n = 0,1,\ldots, m$ is a given positive real finite sequence. With these identifications one finds that the corresponding real solutions of (2) will be continuous polygonal curves whose vertices $(t_n, y(t_n)) \equiv (t_n, y_n)$ have their ordinates, y_n, satisfying the formally symmetric second-order linear difference equation

$$\Delta(C_{n-1}\Delta y_{n-1}) - b_n y_n = 0, \tag{3}$$

for $n = 0,1,\ldots, m-1$, and Δ is the forward difference operator, $\Delta y_n \equiv y_{n+1} - y_n$. So use of (2) now leads one to understand that solutions of (3) should perhaps be interpreted as continuous functions defined on I and not just as the finite sequence $y_{-1}, y_0, \ldots, y_m$ as one may at first sight suspect. That for (3) solutions are to be interpreted as continuous functions, has its historical precedents. For example, M. Bôcher noted in his survey article [6] that a Sturmian theory could be naturally developed for (3) if "solutions" were treated as continuous functions (in fact, the same polygonal curves that were mentioned above). The advantage in using (2) is that a Sturmian theory can be developed for (2) thus simultaneously yielding such a theory for each of (1) and (3).

If in (2) one chooses $\sigma \in C(I)$ (i.e., continuous on I) then (2) is a pure Stieltjes integro-differential equation. If, in addition, $p \in C(I)$ say, then (2) may be integrated once again to yield the Volterra-Stieltjes integral equation

$$y(t) = \alpha + \beta \int_{\gamma}^{t} \frac{ds}{p(s)} + \int_{\gamma}^{t} (t-s)y(s)d\sigma(s) \qquad t \in I$$

Note that (2) also includes equations of mixed type obtained by, say, setting $\sigma \in C^1(I)$ except at a finite number of points or by defining σ to be a C^1-function on a part of I and a step-function elsewhere.

An intensive study of equations of the form (2) was undertaken by F. V. Atkinson [3] in his monograph, (See also the fundamental paper of Krein [39] and the related papers of W.T. Reid [79], [80]).

In order to derive a spectral theory for (2) one needs to use (2) in order to define an operator on some suitable space. To this end, note that if y is a solution of (2) then

$$\frac{d}{dt} \{p(t)y'(t) - \int_{\gamma}^{t} y(s)d\sigma(s)\} = 0 \qquad (4)$$

and conversely if one defines a solution of (4) as a function $y \in AC(I)$ for which $p(t)y'(t) - \int_{\gamma}^{t} y(s)d\sigma(s) \in AC(I)$. We can then recover (2) from (4). On the other hand the left-side of (4) defines a generalized differential expression, viz.

$$\ell[y](t) = -\frac{d}{dt}\{p(t)y'(t) - \int_{\gamma}^{t} y(s)d\sigma(s)\}.$$

and such an expression may then be used to define a linear operator on $L^2(I)$ with due care for domain considerations.

If one wishes to treat boundary problems for Sturm-Liouville equations with a weight-function $r(t) \in L(I)$,

$$-(p(t)y')' + q(t)y = \lambda r(t)y,$$

consideration of the generalized ordinary differential expression

$$\ell[y](t) = -\frac{d}{d\nu(t)}\{p(t)y'(t) - \int_{\gamma}^{t} y(s)d\sigma(s)\}, \quad (5)$$

may be made, where the generalized derivative appearing on the right is, in general, a Radon-Nikodym derivative. The case $r(t) > 0$ corresponds to $\nu(t)$ non-decreasing and the case of unrestricted $r(t)$ corresponds to $\nu(t) \in BV(I)$. In the former case the operator defined by the differential expression is formally symmetric (under suitable domain restrictions) in the weighted Hilbert space $L^2(I,d\nu)$. In the latter case the operator is J-symmetric in a Krein (Pontrjagin) space, since the measure induced by $\nu(t)$ is a signed measure.

Expressions of the form (5) were first considered by W. Feller [68],[69],[70],[71],[72],[73] in the case when $\sigma(t) \equiv$ constant on I, $p(t) \equiv 1$, and ν a given non-decreasing function on I, (cf., also Langer [41]). The more general case $\sigma \in BV(I)$ was treated by I.S. Kac [35],[36],[37] when ν is monotone, cf., [46,p.49].

CHAPTER 1

INTRODUCTION:

In this chapter we shall study the Sturmian theory of Stieltjes integro-differential equations; that is, equations of the form

$$p(t)y'(t) = c + \int_a^t y(s)d\sigma(s) \qquad (1.0.0)$$

defined on a finite interval $I = [a, b]$ and p, σ are real valued right-continuous functions of bounded variation on I and $p(t) > 0$ there.

Historical Background:

The comparison and separation theorems of Sturm comprise what we call the Sturmian theory. Comparison theorems for the scalar equation

$$\big(p(t)y'(t)\big)' - q(t)y(t) = 0 \qquad (1.0.1)$$

were first obtained by Sturm [58, p. 135] in his famous memoir of 1836. In that paper Sturm considered the equations

$$(K_1 y')' - G_1 y = 0 \qquad (1.0.2)$$

$$(K_2 z')' - G_2 z = 0 \tag{1.0.3}$$

on a finite interval and showed that if $0 < K_2 \leqq K_1$, $G_2 \leqq G_1$, equality not holding everywhere on the interval, then between any two zeros of some solution of (1.0.2) there is at least one zero of any solution of (1.0.3). This is the result usually known as the *Sturm-Comparison Theorem*. Sturm's proof depended upon the introduction of a parameter in the coefficients which allowed him to pass continuously from K_1 to K_2 and from G_1 to G_2, as the parameter was increased, and then he studied the location of the zeros of the solutions as the parameter varied. It also depended upon the identity valid for all $t_1, t_2 \in I$,

$$\left[K_2 y z' - K_1 y' z\right]_{t_1}^{t_2} = \int_{t_1}^{t_2} (G_2 - G_1) y z \, dt + \int_{t_1}^{t_2} (K_2 - K_1) y' z' \, dt \tag{1.0.4}$$

which can be obtained by an application of Green's theorem [13, p. 291].

It seems [58, p. 186] that Sturm came to the conclusion of the comparison theorem by first having shown it true for the case of a three-term recurrence relation or second order difference equation though the latter result was not published. A discrete analog of the comparison theorem was published by Fort [21, p.] whose method of proof was, in essence, that of Sturm applied to difference equations instead of differential equations.

In 1909 Picone [48, p. 18] gave by far the simplest proof of the comparison theorem in the continuous case. He made use of the formula

$$\left[\frac{y}{z}(K_2yz' - K_1y'z)\right]_{t_1}^{t_2} = \int_{t_1}^{t_2}(K_2 - K_1)y'^2\,dt + \int_{t_1}^{t_2}(G_2 - G_1)y^2\,dt - \int_{t_1}^{t_2} K_2\left(y' - \frac{z'y}{z}\right)^2 dt \qquad (1.0.5)$$

commonly known as the Picone Identity. The use of (1.0.5) allows an immediate proof of the Sturm Comparison Theorem [33, p. 226]. (cf., also [74]).

One important extension of the comparison theorem was that of Leighton [42, p. 604] who interpreted the theorem in a variational setting: He made use of a "quadratic functional" $Q[y]$ associated with (1.0.2-3) acting on functions $y \in C^1(a, b)$ and $y(a) = y(b) = 0$ (such functions were termed 'admissible'). For such y,

$$Q[y] \overset{df}{=} \int_a^b (K_2y'^2 + G_2y^2)dt \,. \qquad (1.0.6)$$

The main result was that if some non-trivial admissible function y had the property that $Q[y] < 0$ then every real solution of (1.0.3) would have to vanish at some point in (a, b). Swanson [59, p. 3] weakened Leighton's condition $Q[y] < 0$ to $Q[y] \leqq 0$ for $y \not\equiv 0$ reaching the same conclusion provided the solutions were not constant multiples of y.

The *Sturm-Separation theorem* states that the zeros of linearly independent solutions of, say, (1.0.2) interlace or separate one another. A similar result holds for three-term recurrence relations and in fact a more general result is known in the latter case. (See section 2).

In section 1 we shall give an extension of the aforementioned "Leighton-Swanson Theorem" to the class of integral equations (1.0.0) and give, as corollaries, the corresponding continuous and discrete versions of the comparison theorem.

In section 2 we give a proof of the Sturm Separation Theorem for (1.0.0) and give some applications to both differential and difference equations. We conclude this chapter with a study of the Green's function for boundary problems associated with (1.0.0) and its application to the problem of finding an explicit representation for the solution of the non-homogeneous problem. (See section 3).

§1.1 COMPARISON THEOREMS FOR STIELTJES INTEGRO-DIFFERENTIAL EQUATIONS:

Let $p_i(t)$, $\sigma_i(t)$, $i = 1, 2$, be real valued functions of bounded variation over $[a, b]$. We assume that $p_i(t) > 0$, $t \in [a, b]$, $i = 1, 2$, and that all four functions are right-continuous on $[a, b]$ with each possessing a finite number of discontinuities there. (This is for simplicity only. In the following chapters this hypothesis can be omitted, in most theorems, without affecting the conclusions.) We will, in general, assume that all these functions are continous at a, b, and if $b = \infty$, then $\lim \sigma(t)$ exists as $t \to \infty$.

Consider the equations

$$p_1(t)u'(t) = c + \int_a^t u(s)\,d\sigma_1(s) \qquad (1.1.0)$$

$$p_2(t)v'(t) = c' + \int_a^t v(s)\,d\sigma_2(s) \qquad (1.1.1)$$

where by a *solution* of (1.1.0), say, we mean a function $u(t) \in AC[a, b]$ with $p_1(t)u'(t) \in BV(a, b)$ satisfying (1.1.0) at each point $t \in [a, b]$.

Associated with the pair (1.1.0-1) is the quadratic functional $Q[u]$ with domain D_Q

$$D_Q = \{u : u \in AC[a, b],\ p_2u' \in BV(a, b),\ u(a) = u(b) = 0\} \qquad (1.1.2)$$

and where, for $u \in D_Q$,

$$Q[u] = \int_a^b (p_2u'^2dt + u^2d\sigma_2)\ . \qquad (1.1.3)$$

We can now state and prove an extension of the Leighton-Swanson result.

THEOREM 1.1.0:

Let p_i, σ_i, $i = 1, 2$, be defined as above and let

$u \in D_Q$, $u \not\equiv 0$, be such that

$$Q[u] \leqq 0 . \tag{1.1.4}$$

Then every solution of (1.1.1) which is not a constant multiple of $u(t)$ vanishes at least once in (a , b) .

Proof: Assume, on the contrary, that v does not vanish in (a , b) and let $a < s < t < b$. Then $p_2 v'/v \in BV_{loc}(a , b)$ and so

$$\int_s^t u^2 d\left(\frac{p_2 v'}{v}\right) \tag{1.1.5}$$

exists.

Case 1: $v(a) \neq 0$, $v(b) \neq 0$.

For $u \in D_Q$ satisfying (1.1.4),

$$\int_s^t u^2 d\left(\frac{p_2 v'}{v}\right) = \int_s^t u^2 \left[\frac{1}{v}\, dp_2 v' + p_2 v'\, dv^{-1}\right]$$

$$= \int_s^t \left(\frac{u^2}{v}\, dp_2 v' + u^2 p_2 v'\, dv^{-1}\right) \tag{1.1.6}$$

$$= \int_s^t u^2 d\sigma_2 - 2 \int_s^t p_2 \left\{\frac{uv'}{v}\right\}^2 dt \tag{1.1.7}$$

where in passing from (1.1.6) to (1.1.7) we used the equation (1.1.1). Integrating (1.1.5) by parts we find that

$$\int_s^t u^2 d\left(\frac{p_2 v'}{v}\right) = \left[p_2 \frac{v'u^2}{v}\right]_s^t - 2\int_s^t p_2 \frac{v'u'u}{v} . \qquad (1.1.8)$$

Combining (1.1.7), (1.1.8) and adding

$$\int_s^t p_2 u'^2$$

to both sides we obtain,

$$\begin{aligned}\int_s^t (p_2 u'^2 dt + u^2 d\sigma_2) &= \left[p_2 v' \frac{u^2}{v}\right]_s^t + \int_s^t p_2 \left\{\frac{uv'}{v}\right\}^2 \\ &\quad + \int_s^t p_2 u'^2\, dt - 2\int_s^t p_2 v'u' \frac{u}{v} \\ &= \left[p\ v' \frac{u^2}{v}\right]_s^t + \int_s^t p_2 \left\{u' - \frac{uv'}{v}\right\}^2 \\ &= \left[p_2 v' \frac{u^2}{v}\right]_s^t + \int_s^t p_2 v^2 \left\{\frac{u}{v}\right\}'^2\end{aligned} \qquad (1.1.9)$$

for $a < s < t < b$.

Hence if we let $s \to a+0$, $t \to b-0$ in (1.1.9) we obtain, since $v(a)$, $v(b) \neq 0$,

$$Q[u] = \int_a^b p_2 v^2 \left\{\frac{u}{v}\right\}'^2 \geqq 0 . \qquad (1.1.10)$$

The hypothesis on u implies $Q[u] = 0$ but since $v \not\equiv 0$, we must have $\left\{\frac{u}{v}\right\}' = 0$ or that u is a multiple of v on $[a, b]$ which we excluded. This contradiction shows that v

must vanish at least once in (a, b) .

<u>Case 2</u>: $v(a) = v(b) = 0$.

To settle this case it suffices to show that in (1.1.9),

$$\lim_{t \to b-0} \frac{u^2(t)p_2(t)v'(t)}{v(t)} = 0 \tag{1.1.11}$$

and

$$\lim_{s \to a+0} \frac{u^2(s)p_2(s)v'(s)}{v(s)} = 0 . \tag{1.1.12}$$

It is possible to show that solutions to the initial value problem (1.1.1), $v(a) = c_1$, $p_2(a)v'(a) = c_2$ are unique: See Appendix I and [3, p. 341]. Thus since $v(a) = 0$, $v'(a) \neq 0$. (The prime here usually represents a right-derivative which is an ordinary (two-sided) derivative if σ_2 is continuous at the point in question.) Similarly [3, p. 348], since $v(b) = 0$, $p_2(b)v'(b) \neq 0$. Hence

$$\lim_{s \to a+0} \frac{u^2(s)p_2(s)v'(s)}{v(s)} = p_2(a)v'(a) \lim_{s \to a+0} \frac{u^2(s)}{v(s)} \tag{1.1.13}$$

provided the latter limit exists. The hypothesis on σ_2 implies that it is continuous in some right-neighborhood of a . Thus $p_2(t)v'(t)$ is continuous in such a neighborhood. Similarly $p_2(t)$ is continuous in some, possibly different, right-neighborhood of a . Hence $v'(t)$ is continuous (i.e.

is an ordinary derivative) in some right-neighborhood $(a, a+\delta)$, $\delta > 0$.

In the same way it can be shown that $u'(t)$ is an ordinary derivative in $(a, a+\eta)$, $\eta > 0$. Thus in $(a, a+\eta)$, $(u^2(t))' = 2u(t)u'(t)$. Since $u, v \in AC[a, b]$, we can apply L'Hôpital's theorem to the limit in the right of (1.1.13) to obtain

$$\lim_{s\to a+0} \frac{u^2(s)}{v(s)} = \lim_{s\to a+0} \frac{2u(s)u'(s)}{v'(s)}$$

$$= 0$$

since, as we saw above, $v'(a) \neq 0$. Hence the limit (1.1.12) exists and is zero.

Similarly it can be shown that (1.1.11) holds. Combining (1.1.11), (1.1.12) and letting $s \to a+0$, $t \to b-0$ in (1.1.9) we obtain (1.1.10) again and thus derive a contradiction.

<u>Case 3</u>: $v(a) = 0$, $v(b) \neq 0$ or $v(a) \neq 0$, $v(b) = 0$.

This case is easily disposed of as it is simply a combination of Cases 1 and 2 leading to (1.1.10) via (1.1.9) and (1.1.11-12). This proves the theorem.

Associated with (1.1.0) is the quadratic functional $Q'[u]$ with domain $D_{Q'}$

$$D_{Q'} = \{u : u \in AC[a, b] , p_1 u' \in BV(a, b) , u(a) = u(b) = 0\}$$

(1.1.14)

and

$$Q'[u] = \int_a^b (p_1 u'^2 dt + u^2 d\sigma_1) . \qquad (1.1.15)$$

<u>COROLLARY 1.1.0</u>: (Swanson [59, p. 4], Leighton [42, p. 605, Cor. 1]).

Let u be a non-trivial solution of (1.1.0) with $u(a) = u(b) = 0$.

Then every solution $v(t)$ of (1.1.1) which is not a constant multiple of u must vanish at least once in (a, b) provided

$$\int_a^b \{(p_1 - p_2)u'^2 dt + u^2 d(\sigma_1 - \sigma_2)\} \geqq 0 . \qquad (1.1.16)$$

<u>Proof</u>: Let u be a solution of (1.1.0), $u(a) = u(b) = 0$. Then

$$\int_a^b u\, d(p_1 u') = [u p_1 u']_a^b - \int_a^b p_1 u'^2\, dt . \qquad (1.1.17)$$

Using the equation (1.1.0) in the left-side of (1.1.17) we find that

$$\begin{aligned} Q'[u] &= \int_a^b (p_1 u'^2 dt + u^2 d\sigma_1) \\ &= [u p_1 u']_a^b \\ &= 0 . \end{aligned} \qquad (1.1.18)$$

(1.1.16) now says that $Q'[u] - Q[u] \geqq 0$ or, because of (1.1.18),

$$Q[u] \leqq 0 . \tag{1.1.19}$$

Since u is not a constant multiple of v, Theorem 1.1.0 applies and hence $v(t)$ vanishes at least once in (a , b). Swanson's extension [59, p. 4] of Leighton's Theorem [42] is obtained by setting

$$\sigma_i(t) = \int_a^t q_i \, ds , \quad t \in [a , b] , \quad i = 1 , 2 , \tag{1.1.20}$$

in (1.1.0-1) and in (1.1.16).

<u>COROLLARY 1.1.1:</u> (Sturm Comparison Theorem)

Let p_i, $q_i \in C[a , b]$, $i = 1 , 2$ and suppose that $p_1(t) \geqq p_2(t) > 0$, $q_1(t) \geqq q_2(t)$.

If

$$(p_1 u')' - q_1 u = 0 \tag{1.1.21}$$

$$(p_2 v')' - q_2 v = 0 \tag{1.1.22}$$

and $u(a) = u(b) = 0$, then there is at least one $c \in (a , b)$ for which $v(c) = 0$ whenever v is a solution of (1.1.22) which is not a constant multiple of u.

Proof: Let $\sigma_i(t)$ be defined as in (1.1.20). The result now follows from Corollary 1.1.0 on account that $\sigma_1(t) - \sigma_2(t)$ is non-decreasing on $[a, b]$ by the above hypothesis.

We now interpret these results for a three-term recurrence relation. Let

$$t_{-1} = a < t_0 < t_1 < \cdots < t_{m-1} < t_m = b \qquad (1.1.23)$$

be a fixed partition of the interval $[a, b]$ and let $c_{-1}, c_0, c_1, \ldots, c_{m-1}$ be a given positive real sequence. Let $b_0, b_1, \ldots, b_{m-1}$ be an arbitrary real sequence and define a function $p(t)$ on $[a, b]$ by setting

$$p(t) = c_{n-1}(t_n - t_{n-1}) \quad \text{if} \quad t \in [t_{n-1}, t_n) \qquad (1.1.24)$$

for $n = 0, 1, 2, \ldots, m$. Then $p(t)$ is a positive right-continuous function of bounded variation with jumps, if any, at the $\{t_i\}$.

Now define $\sigma(t)$ on $[a, b]$ by requiring that it be a right-continuous step-function with jumps at the $\{t_i\}$ of magnitude

$$\sigma(t_n) - \sigma(t_n - 0) = -b_n \qquad (1.1.25)$$

where $n = 0, 1, \ldots, m-1$.

With $p(t)$, $\sigma(t)$ as defined above consider (1.0.0). On

$[a, t_0)$, $\sigma(t)$ = constant, hence,

$$\int_a^t y\,d\sigma \equiv 0 \qquad t \in [a, t_0) , \tag{1.1.26}$$

and so (1.1.0) implies that

$$p(t)y'(t) = c = p(a)y'(a) \qquad t \in [a, t_0) . \tag{1.1.27}$$

But $p(t) = p(a)$ on $[a, t_0)$ because $p(t)$ is also a step-function, hence (1.1.27) implies that

$$y'(t) = y'(a) \qquad t \in [a, t_0) \tag{1.1.28}$$

In fact, letting $y(t_n) = y_n$, $n = -1, 0, 1, \ldots, m$, then (1.1.27) gives

$$y'(t) = \frac{y(t_0) - y(a)}{t_0 - a}$$

$$= \frac{y_0 - y_{-1}}{t_0 - t_{-1}} \qquad t \in [a, t_0) . \tag{1.1.29}$$

Hence

$$p(t)y'(t) = p(a)y'(a)$$

$$= c_{-1}(t_0 - t_{-1}) \cdot \frac{(y_0 - y_{-1})}{t_0 - t_{-1}}$$

$$= c_{-1}(y_0 - y_{-1}) \qquad t \in [a, t_0) . \tag{1.1.30}$$

Now let $t \in [t_{n-1}, t_n)$, $1 \leq n \leq m$. When $n = 0$, we know from (1.1.30) that $p(t)y'(t) = c_{-1}(y_0 - y_{-1})$ for $t \in [a, t_0)$. Thus

$$\begin{aligned} p(t)y'(t) &= p(a)y'(a) + \int_a^t y\,d\sigma \\ &= p(a)y'(a) + \sum_{i=0}^{n-1} \int_{t_{i-1}}^{t_i} y\,d\sigma + \int_{t_{n-1}}^{t} y\,d\sigma \\ &= p(a)y'(a) + \sum_{i=0}^{n-1} \int_{t_i-0}^{t_i+0} y\,d\sigma + \int_{t_{n-1}+0}^{t} y\,d\sigma \\ &= p(a)y'(a) + \sum_{i=0}^{n-1} y(t_i)\big(\sigma(t_i) - \sigma(t_i - 0)\big) + 0 \end{aligned}$$

since σ is constant on $[t_{n-1}, t_n)$. Hence $p(t)y'(t)$ is constant on $[t_{n-1}, t_n)$ and since $p(t)$ satisfies (1.1.24) there, $y'(t)$ is also constant so that

$$y'(t) = \frac{y_n - y_{n-1}}{t_n - t_{n-1}} \qquad t \in [t_{n-1}, t_n) .$$

Consequently,

$$p(t)y'(t) = c_{n-1}(y_n - y_{n-1}) \qquad t \in [t_{n-1}, t_n) . \tag{1.1.31}$$

This is true for each n in the range considered. If $t \in [t_n, t_{n+1})$, (1.1.0) gives

$$p(t)y'(t) = p(t_n - 0)y'(t_n - 0) + \int_{t_n-0}^{t} y\,d\sigma$$

$$= p(t_n - 0)y'(t_n - 0) + \int_{t_n-0}^{t_n+0} y\,d\sigma$$

$$= c_{n-1}(y_n - y_{n-1}) + y(t_n)\big(\sigma(t_n) - \sigma(t_n - 0)\big) \qquad (1.1.32)$$

$$= c_{n-1}(y_n - y_{n-1}) + y_n(-b_n) \qquad (1.1.33)$$

where we have used (1.1.31) and (1.1.25) in obtaining (1.1.32), (1.1.33) respectively.

By (1.1.31) we find that $p(t)y'(t) = c_n(y_{n+1} - y_n)$ if $t \in [t_n, t_{n+1})$. Combining this with (1.1.33) we obtain

$$c_n(y_{n+1} - y_n) = c_{n-1}(y_n - y_{n-1}) - b_n y_n \qquad (1.1.34)$$

or

$$c_n y_{n+1} + c_{n-1} y_{n-1} - (c_n + c_{n-1} - b_n) y_n = 0 \qquad (1.1.35)$$

which is equivalent to

$$\Delta(c_{n-1}\Delta y_{n-1}) + b_n y_n = 0 \qquad (1.1.36)$$

where Δ represents the forward difference operator, $\Delta y_n = y_{n+1} - y_n$.

Summarizing then, we see that when $p(t)$, $\sigma(t)$ are defined as in (1.1.24-5) respectively, the Stieltjes integro-differential equation (1.1.0) has solutions which are polygonal curves whose "vertices" are the points (t_n, y_n) and the y_n satisfy the three-term recurrence relation (1.1.35) or the second-order difference equation (1.1.36) for $n = 0, 1, 2, \ldots, m-1$.

An argument similar to the one above shows that if, instead of (1.1.25), we require

$$\sigma(t_n) - \sigma(t_n - 0) = b_n - c_n - c_{n-1}, \qquad (1.1.37)$$

then (1.1.0), with the same p as in (1.1.24), will give rise to the recurrence relation

$$c_n y_{n+1} + c_{n-1} y_{n-1} - b_n y_n = 0 \qquad (1.1.38)$$

where $n = 0, 1, 2, \ldots, m-1$.

The initial conditions

$$y(a) = \alpha \qquad (1.1.39)$$

$$p(a)y'(a) = \beta \qquad (1.1.40)$$

associated with (1.1.0) become, in the case of a recurrence relation,

$$y_{-1} = \alpha \qquad (1.1.41)$$

$$c_{-1}(y_0 - y_{-1}) = \beta \qquad (1.1.42)$$

on account of (1.1.30).

A fundamental solution, i.e. one in which $\alpha = 0$, $\beta = 1$, will then become

$$y_{-1} = 0 \tag{1.1.43}$$

$$y_0 = \frac{1}{c_{-1}} \tag{1.1.44}$$

in the case of a recurrence relation (1.1.38). (In this respect see [3, p. 97]).

With the t_n defined as in (1.1.23) we suppose given two arbitrary real finite sequences b_n , q_n , $n = 0, 1, \ldots, m-1$ and two positive sequences c_n , r_n , $n = -1, 0, 1, \ldots, m-1$. Let $\sigma_i(t)$, $i = 1, 2$, be step-functions on $[a, b]$ with saltus

$$\sigma_1(t_n) - \sigma_1(t_n - 0) = b_n \tag{1.1.45}$$

$$\sigma_2(t_n) - \sigma_2(t_n - 0) = q_n \tag{1.1.46}$$

$n = 0, 1, \ldots, m-1$.

Let $p_i(t)$, $i = 1, 2$, be defined by

$$p_1(t) = c_{n-1}(t_n - t_{n-1}) \qquad t \in [t_{n-1}, t_n) \tag{1.1.47}$$

$$p_2(t) = r_{n-1}(t_n - t_{n-1}) \qquad t \in [t_{n-1}, t_n) \tag{1.1.48}$$

where $n = 0, 1, \ldots, m-1$. Then the $p_i(t) > 0$ for $i = 1, 2$, and, along with the $\sigma_i(t)$, $i = 1, 2$, are right continuous and of bounded variation on $[a, b]$.

Consider now (1.1.0-1) with the above choice of p_i, σ_i, $i = 1, 2$. The solutions of (1.1.0-1) evaluated at the points t_i will then satisfy the recurrence relations

$$c_n u_{n+1} + c_{n-1} u_{n-1} - (c_n + c_{n-1} + b_n) u_n = 0$$

and

$$r_n v_{n+1} + r_{n-1} v_{n-1} - (r_n + r_{n-1} + q_n) v_n = 0$$

where $n = 0, 1, \ldots, m-1$ respectively. The latter are equivalent to

$$\Delta(c_{n-1} \Delta u_{n-1}) - b_n u_n = 0 \tag{1.1.49}$$

$$\Delta(r_{n-1} \Delta v_{n-1}) - q_n v_n = 0 \tag{1.1.50}$$

$n = 0, 1, \ldots, m-1$. We can now state a discrete analog of the Sturm comparison theorem one form of which was proven by Fort [21, p.].

COROLLARY 1.1.2:

Let $c_n \geqq r_n > 0$ and $b_n \geqq q_n$ for $n = 0, 1, \ldots, m-1$, equality not holding for every n. If $u_{-1} = u_m = 0$ and

$$\Delta(c_{n-1} \Delta u_{n-1}) - b_n u_n = 0 \tag{1.1.51}$$

then there is at least one node of

$$\Delta(r_{n-1}\Delta v_{n-1}) - q_n v_n = 0 \qquad (1.1.52)$$

in (a, b).

REMARK:

We note that the condition $u_{-1} = u_m = 0$ is equivalent to $u(a) = u(b) = 0$ when u is considered a solution of (1.1.0).

By a *node* we mean a point on the abscissa where the "polygonal curve" defined by the finite sequence v_n crosses the axis.

Proof: The condition $c_n \geqq r_n > 0$ along with (1.1.47-8) implies that $p_1(t) \geqq p_2(t) > 0$. Moreover, since $b_n \geqq q_n$ we find from (1.1.45-6) that

$$\sigma_1(t_n) - \sigma_2(t_n) \geqq \sigma_1(t_n - 0) - \sigma_2(t_n - 0). \qquad (1.1.53)$$

Since σ_1, σ_2 are step-functions on (t_{n-1}, t_n) for each n, (1.1.53) implies that $\sigma_1(t) - \sigma_2(t)$ is non-decreasing for $t \in [a, b]$. This, along with the above Remark, shows that Corollary 1.1.0 is applicable and hence the equation (1.1.1) has at least one zero in (a, b) which is equivalent to the required conclusion.

Note: In general, a comparison theorem for equations of the form

$$c_n y_{n+1} + c_{n-1} y_{n-1} - b_n y_n = 0 \qquad (1.1.54)$$

$$r_n z_{n+1} + r_{n-1} z_{n-1} - q_n z_n = 0 \qquad (1.1.55)$$

under the assumptions $c_n \geqq r_n$, $b_n \geqq q_n$ is not available. For example let $c_n = c > 0$ for all n, and $b_n = 3c$, $r_n = q_n = c/2$ for each n, $n = 0, 1, \ldots, m-1$. We see then that $c_n > r_n$, $b_n > q_n$ but a simple computation shows that, in this case, (1.1.54) has no nodes eventually while (1.1.55) will have nodes for large n.

§1.2 SEPARATION THEOREMS:

In this section we prove the classical Sturm separation theorem, on the separation of zeros of linearly independent solutions, as a consequence of the results in section 1. In the case of finite differences this result was also probably known to Sturm [58, p. 186] as one can gather from the remarks at the end of his memoir.

If $c_n > 0$ for all n, then the nodes of solutions of

$$c_n y_{n+1} + c_{n-1} y_{n-1} - b_n y_n = 0 \qquad (1.2.0)$$

separate one another if these are linearly independent. (The proof of this result will follow below.)

The Sturm separation theorem is not valid in general in the case of a general three-term recurrence relation

$$P_n y_{n+1} + Q_n y_n + R_n y_{n-1} = 0 . \qquad (1.2.1)$$

Bôcher [6, p. 176] points out that the separation property for solutions of (1.2.1) holds if

$$P_n R_n > 0 \qquad (1.2.2)$$

for all n in the range considered. The result is however false, in general, if (1.2.2) fails. He gives [6, p. 177] as an example the case where $P_n = 1$, $Q_n = R_n = -1$ for all n . The nodes of the linearly independent solutions corresponding to the initial values $y_{-1} = 0$, $y_0 = 1$ and $y_{-1} = -10$, $y_0 = 6$ do not separate one another.

One proof of the separation property of (1.2.1) under the hypothesis (1.2.2) was given by Moulton [45, p. 137]. We note that the condition (1.2.2) is the analog of the condition $p(t) > 0$ for the equation

$$p(t)y'' + q(t)y' + r(t)y = 0 . \qquad (1.2.3)$$

If $p(t) > 0$ then the zeros of linearly independent solutions of (1.2.3) separate one another. (One way of seeing this is that (1.2.3) can then be transformed into an equation of the form

$$\big(P(t)y'\big)' + Q(t)y = 0 \tag{1.2.4}$$

where $P(t) > 0$ and the result follows from the separation property of the zeros of (1.2.4).)

<u>THEOREM 1.2.0</u>:

The zeros of linearly independent solutions of

$$p(t)y'(t) = c + \int_a^t y(s)\,d\sigma(s) \tag{1.2.5}$$

separate one another.

<u>Proof</u>: (1.2.5) has two linearly independent solutions u, v which generate the solution space [3, p. 348]. If we now set $\sigma_1 = \sigma_2$ and $p_1 = p_2$ in (1.1.16) we can apply Corollary 1.1.0 to, say, u when u vanishes at two consecutive points to find that v must vanish in between since v is not a constant multiple of u.

In particular if $\sigma \in C'(a, b)$ we immediately obtain the classical Sturm separation theorem.

<u>COROLLARY 1.2.0</u>:

If $\sigma \in C'(a, b)$ and $\sigma'(t) = q(t)$, $t \in [a, b]$ then the zeros of linearly independent solutions of

$$(p(t)y')' - q(t)y = 0 \tag{1.2.6}$$

separate each other.

Porter [49, p. 55] showed that two linearly independent solutions of (1.2.0) generate the solution space and considered the limiting process which takes a difference equation to a differential equation.

Defining σ, p as in (1.1.24-25) we obtain the discrete analog

<u>COROLLARY 1.2.1</u>:

If $c_n > 0$, $n = -1, 0, \ldots, m-1$ and b_n , $n = 0, 1, \ldots, m-1$ is any sequence and

$$\Delta(c_{n-1}\Delta y_{n-1}) - b_n y_n = 0 \tag{1.2.7}$$

then the nodes of linearly independent solutions separate one another.

As an application of Corollary 1.2.1 to the recurrence relation (1.2.1) we state the following [45, p. 137].

<u>COROLLARY 1.2.2</u>:

Let P_n, Q_n, R_n be real finite sequences and

$$P_n y_{n+1} + Q_n y_n + R_n y_{n-1} = 0 \qquad (1.2.8)$$

for $n = 0, 1, \ldots, m-1$.

If

$$P_n R_n > 0 \qquad n = 0, 1, \ldots, m-1 \qquad (1.2.9)$$

then the nodes of linearly independent solutions of (1.2.8) separate each other.

<u>Proof</u>: The idea is to show that (1.2.8) under the hypothesis (1.2.9) can be brought into the form (1.2.7) after which we simply apply Corollary 1.2.1.

Let $c_{-1} > 0$ and consider the recurrence relation

$$c_n = \frac{P_n}{R_n} c_{n-1} \qquad n = 0, 1, \ldots, m-1 . \qquad (1.2.10)$$

(1.2.9) implies that $c_n > 0$ for $n = 0, 1, \ldots, m-1$ since $c_{-1} > 0$. If we now set

$$b_n = -c_n - c_{n-1} - \frac{c_n Q_n}{P_n} \qquad (1.2.11)$$

for $n = 0, 1, \ldots, m-1$, then a simple computation shows that with the substitutions (1.2.10-11), (1.2.7) reduces to the three-term recurrence relation (1.2.8). Hence the result follows.

§1.3. The GREEN'S FUNCTION:

In Appendix I to this work we have shown the existence of a Green's function for the inhomogeneous problem

$$\psi(t) = \alpha + \beta \int_a^t \frac{1}{p} + \int_a^t \frac{1}{p(s)} \int_a^s \psi d\sigma \, ds + \int_a^t \frac{f}{p} \tag{1.3.0}$$

$$U_1\psi = U_2\psi = 0 \tag{1.3.1}$$

where

$$U_i\psi = \sum_{j=1}^{2} \left\{ M_{ij}\psi^{(j-1)}(a) + N_{ij}\, p(b)\psi^{(j-1)}(b) \right\} \qquad i = 1 , 2 , \tag{1.3.2}$$

and the M_{ij} , N_{ij} are real constants, under the hypothesis that the homogeneous problem (with $f = 0$) and the boundary conditions (1.3.1) is *incompatible*. (By this we mean that the homogeneous equation with homogeneous boundary conditions has only the zero solution.) If $f = 0$ in (1.3.0) then the resulting integral equation is of the form (1.0.0).

If $\sigma \in C'(a , b)$ and $p(t)$ is positive and continuous on $[a , b]$ then (1.3.0) with $f = 0$ reduces to (1.2.6) with $\sigma' = q$. In this case the "derivative" appearing in (1.0.0) is continuous and the Green's function reduces to the usual one. (See Appendix I, p. 278.)

On the other hand if $p(t)$, $\sigma(t)$ are step-functions then (1.3.0) with $f = 0$ can be made to include three-term recurrence relations. In this case and, more generally, for difference equations of higher order the Green's function seems to have been first constructed by Bôcher [5, p. 83]. Another treatment was given by Atkinson [3, p. 148].

We showed in Appendix I that if (1.3.0-1) with $f = 0$ is incompatible then the unique solution of (1.3.0-1) is given by

$$\psi(x) = \int_a^b G(x, t)\,df(t) \qquad (1.3.3)$$

for $x \in [a, b]$. In the particular case when $p(t) = 1$, $t \in [a, b]$ and $f(t)$ is a step-function with jumps at the same points where $\sigma(t)$ has its jumps and if we denote by

$$f_i = f(t_i) - f(t_i - 0) \qquad (1.3.4)$$

where, as usual, the t_i represent the jump points of f, then a simple computation shows that

$$\psi_n \equiv \psi(t_n) = \int_a^b G(t_n, t)\,df(t)$$

$$= \sum_{i=0}^{m-1} G(t_n, t_i) \cdot \big(f(t_i) - f(t_i - 0)\big)$$

(1.3.5)

and if we write $G_{ni} \equiv G(t_n , t_i)$, $0 \leqq n$, $i \leqq m-1$ we find that

$$\psi_n = \sum_{i=0}^{m-1} G_{ni} f_i \ . \tag{1.3.6}$$

This ψ_n then represents the solution to the corresponding inhomogeneous difference boundary problem. Usually (1.3.6) is derived directly using methods of finite differences. (See for example [3, p. 149] and [5, p. 84].) For further details see Appendix I, section I.4.

We note that when $p(t)$, $\sigma(t)$ are continuous functions of bounded variation on $[a , b]$ then the derivative appearing in (1.0.0) is continuous everywhere and so, from Appendix I, the discontinuity in the first derivative of the Green's function is given by

$$G_x(t+0 , t) - G_x(t-0 , t) = \frac{1}{p(t)} \tag{1.3.7}$$

which is the usual measure of discontinuity of the Green's function associated with a second-order linear differential equation of the form (1.2.6).

CHAPTER 2

INTRODUCTION:

There is a very extensive literature dealing with the subject of oscillation and non-oscillation of real second order differential equations on a half-axis (see, for example, [59]). On the other hand there is little known about establishing criteria for the oscillatory and non-oscillatory behaviour of solutions of difference equations. In the particular case of three-term recurrence relations some results can be found in [23, pp. 126-128] and more recently in [32, p. 425]. Other results are more or less scattered: [21], [12], [20].

In this chapter we shall be concerned with obtaining some oscillation and non-oscillation criteria for linear and non-linear Stieltjes integral equations on a half-axis. It will be noted that if one makes an hypothesis on the integral of the potential q in

$$y'' - q(t)y = 0 \qquad t \in [a, \infty) \tag{2.0.0}$$

which will guarantee the existence of oscillatory or non-oscillatory solutions, then a certain discrete analog will exist for a three-term recurrence relation.

In section 1 we give some non-oscillation criteria for Stieltjes integral equations and their applications to second order difference equations. In section 2 we give some results on the oscillatory behaviour of solutions and in section 3 we extend a result of Butler [8, p. 75] and state a necessary and sufficient condition which guarantees that all continuable solutions of a non-linear equation are oscillatory. As a corollary we shall obtain the discrete analog of Atkinson's theorem [2, p. 643]. Various examples are included which should help visualize the theorems stated.

§2.1 NON-OSCILLATION CRITERIA FOR LINEAR VOLTERRA-STIELTJES INTEGRAL EQUATIONS:

In the following, we shall usually be considering equations of the form

$$y'(t) = c + \int_a^t y(s)d\sigma(s) \qquad t \in [a,\infty), \tag{2.1.0}$$

where σ is a right-continuous function locally of bounded variation on $[a,\infty)$. Because of the applications we shall assume, in addition, that the number of discontinuities of σ remains finite in finite intervals. The theorems proved here can also be extended to equations of the form

$$p(t)y'(t) = c + \int_a^t y(s)d\sigma(s) \tag{2.1.1}$$

in the case when $p(t) > 0$,

$$\int_a^\infty \frac{1}{p} = \infty \tag{2.1.2}$$

p satisfying the usual conditions stated in Chapter 1. For every equation of the form (2.1.1), where p satisfies (2.1.2), can be transformed into an equation of the form (2.1.0) by the change of independent variable

$$t \mapsto \tau(t) = \int_a^t \frac{1}{p} \tag{2.1.3}$$

which will take $[a, \infty)$ into $[0, \infty)$. (See Appendix I, equation (I.3.14).)

DEFINITION 2.1.1:

A solution of (2.1.0) is said to be *oscillatory* if it has, to the right of a , an infinite number of zeros and is *non-oscillatory* if there is some $t_0 \in \mathbb{R}$ such that it has no zeros when $t \geq t_0$.

From the Sturm separation theorem, Theorem 1.2.0, we see that if one solution is oscillatory (non-oscillatory) then all solutions are oscillatory (non-oscillatory).

Equation (2.1.0) is said to be *oscillatory (non-oscillatory)* if all of its solutions are *oscillatory (non-oscillatory)*.

Unless otherwise stated we shall, in the following,

assume that $\sigma(t)$, appearing in (2.1.0), has a limit at ∞ , i.e.

$$\lim_{t\to\infty} \sigma(t) \tag{2.1.4}$$

exists and is finite. Denoting this limit by $\sigma(\infty)$ we can assume it is zero (for if we let $\tau(t) = \sigma(t) - \sigma(\infty)$ then τ has the same properties as σ and $\tau(\infty) = 0$. Moreover, (2.1.0) remains unchanged if σ is replaced by τ) .

The first result is an extension of a well-known theorem of Hille [31, p. 243] which relates the non-oscillatory behaviour of (2.1.0) to the existence of solutions of a certain non-linear integral equation.

<u>THEOREM 2.1.1</u>:

Let σ be right-continuous and locally of bounded variation satisfying (2.1.4) with $\sigma(\infty) = 0$. Then a necessary and sufficient condition for (2.1.0) to be non-oscillatory is that the integral equation

$$v(t) = \sigma(t) + \int_t^{\infty} v^2(s)ds \tag{2.1.5a}$$

have a solution, for sufficiently large t , which is square integrable at infinity (cf., [80]).

<u>Proof</u>: To show that the condition is sufficient assume that (2.1.4) has a solution $v \in L^2(t_0, \infty)$, some t_0 . (2.1.4) then implies that $v(t)$ is right-continuous, locally of

bounded variation and $v(\infty) = 0$.

Put

$$y(t) = \exp \int_a^t v(s)ds . \qquad (2.1.5b)$$

Then $y(t)$ is locally absolutely continuous and so

$$y'(t) = v(t)\exp \int_{t_0}^t v(s)ds \qquad (2.1.6)$$

everywhere, as a two-sided derivative, except possibly the jump points of $v(t)$ which are the same as those of $\sigma(t)$. Letting $h > 0$, t arbitrary,

$$\frac{y(t+h) - y(t)}{h} = y(t) \cdot \left\{\frac{\exp\left(\left(\int_t^{t+h} v(s)ds\right) - 1\right)}{h}\right\} . \qquad (2.1.7)$$

Now

$$\frac{1}{h}\left\{\exp \int_t^{t+h} v - 1\right\} = \frac{1}{h}\left\{\int_t^{t+h} v + \frac{1}{2!}\left(\int_t^{t+h} v\right)^2 + \cdots\right\} \qquad (2.1.8)$$

for each $h > 0$, fixed t . Hence we can let $h \to 0^+$ and use Theorem H of Appendix I to find that

$$\lim_{h\to 0^+} \frac{1}{h} \int_t^{t+h} v(s)ds = v(t) \qquad (2.1.9)$$

while the other terms are zero by virtue of (2.1.9) and the continuity of the integrals.

Hence letting $h \to 0^+$ in (2.1.7) we obtain, from above,

$$y'(t) = y(t)v(t) \tag{2.1.10}$$

where the derivative is in general understood as a right-derivative which is locally of bounded variation. Thus if $t \geqq t_0$,

$$\begin{aligned}
y'(t) - y'(t_0) &= \int_{t_0}^{t} dy'(s) \\
&= \int_{t_0}^{t} d\big(y(s)v(s)\big) \\
&= \int_{t_0}^{t} v\,dy + \int_{t_0}^{t} y\,dv \\
&= \int_{t_0}^{t} v\,dy + \int_{t_0}^{t} y\,d\sigma - \int_{t_0}^{t} yv^2
\end{aligned}$$

where we have by equation (2.1.10),

$$= \int_{t_0}^{t} (v\,dy - vy'ds) + \int_{t_0}^{t} y\,d\sigma\,. \tag{2.1.11}$$

Theorem K of Appendix I now implies that the first integral in (2.1.11) vanishes for all t and hence

$$y'(t) = y'(t_0) + \int_{t_0}^{t} y\,d\sigma \qquad t \geqq t_0$$

so that $y(t)$ is a positive solution of the above integral equation. This implies that (2.1.0) has a positive solution for $t \geqq t_0$ and hence is non-oscillatory.

To prove the necessity we suppose that (2.1.0) has a non-oscillatory solution $y(t)$ which we can suppose is positive for $t \geqq t_0$.

For $t \geqq t_0$ we set

$$v(t) = \frac{y'(t)}{y(t)} . \tag{2.1.12}$$

Then $v(t)$ is locally of bounded variation on $[t_0, \infty)$ and is right-continuous.

Hence, for $t \geqq t_0$,

$$v(t) - v(t_0) = \int_{t_0}^{t} dv(s)$$

$$= \int_{t_0}^{t} \frac{1}{y(s)}\,dy'(s) - \int_{t_0}^{t} \left(\frac{y'}{y}\right)^2 ds$$

$$= \int_{t_0}^{t} d\sigma(s) - \int_{t_0}^{t} v^2(s)ds .$$

Hence

$$v(t) = \sigma(t) - \sigma(t_0) + v(t_0) - \int_{t_0}^{t} v^2 \qquad (2.1.13)$$

for $t \geqq t_0$. Since $\sigma(t)$ has a limit at ∞ (2.1.13) shows that the same must be true of $v(t)$.

Suppose, if possible, that $v(\infty) = \alpha \neq 0$. Then v cannot be square-integrable at ∞ and so (2.1.13) implies that

$$\lim_{t\to\infty} v(t) = -\infty . \qquad (2.1.14)$$

Hence $y'(t) < 0$ for $t \geqq t_1$ because of (2.1.12). Moreover there is a t_2 such that when $T \geqq t_2$,

$$v(t_2) + \sigma(T) - \sigma(t_2) < -1 . \qquad (2.1.15)$$

If we let $t_3 = \max\{t_0 , t_1 , t_2\}$ then using (2.1.15) in (2.1.13) with t , t_0 replaced by T , t_3 respectively, we obtain

$$v(T) \leqq -1 + \int_{t_3}^{T} \left|\frac{y'(s)}{y(s)}\right| v(s)ds \qquad (2.1.16)$$

whenever $T \geqq t_3$. We now use Gronwall's inequality [9, p. 37, Exercise 1] in (2.1.16) to obtain

$$v(T) \leqq -1 - \int_{t_3}^{T} \left|\frac{y'(s)}{y(s)}\right| \exp\left\{\int_s^T \left|\frac{y'}{y}\right|\right\} ds$$

$$= -\exp \int_{t_3}^{T} \left|\frac{y'(s)}{y(s)}\right| ds$$

$$= -\frac{y(t_3)}{y(T)} .$$

Thus, by (2.1.12),

$$y'(T) \leqq -y(t_3) \qquad (2.1.17)$$

for all $T \geqq t_3$. (2.1.17) implies that $y(t)$ cannot remain positive which is a contradiction. Hence $v(\infty) = 0$. We can now rewrite (2.1.13) as

$$v(T) - v(t) = \sigma(T) - \sigma(t) - \int_t^T v^2 \qquad (2.1.18)$$

where $T \geq t \geq t_3$. Now letting $T \to \infty$ in (2.1.18) we find that $v(t)$ satisfies (2.1.4) for $t \geqq t_3$. This completes the proof.

<u>THEOREM 2.1.2</u>:

Let σ_1 , σ_2 be right-continuous functions locally of bounded variation on $[a , \infty)$ satisfying (2.1.4) with $\sigma_i(\infty) = 0$, $i = 1 , 2$.

Assume that

$$|\sigma_1(t)| \geqq |\sigma_2(t)| \qquad t \geqq t_0 . \tag{2.1.19}$$

If

$$v_1(t) = |\sigma_1(t)| + \int_t^\infty v_1^2 \, ds \tag{2.1.20}$$

has a solution for $t \geqq t_0$ then

$$v(t) = \pm\sigma_2(t) + \int_t^\infty v^2 \, ds \tag{2.1.21}$$

has a solution for $t \geqq t_0$.

Proof: We shall make use of the Schauder fixed point theorem (Appendix II, Theorem 2.1.1). With the Banach space $L^2(t_0, \infty)$ and the usual norm we consider the subset

$$X = \{v \in L^2(t_0, \infty) : |v(t)| \leqq v_1(t) , t \geqq t_0\} \tag{2.1.22}$$

where $v_1(t)$ is as in (2.1.20).

For $v \in X$ we define an operator T on X by

$$(Tv)(t) = \sigma_2(t) + \int_t^\infty v^2 \, ds \qquad t \geqq t_0 . \tag{2.1.23}$$

If $\alpha \in [0, 1]$ and $x , y \in X$,

$$|\alpha x + (1-\alpha)y| \leqq \alpha|x| + (1-\alpha)|y|$$
$$\leqq \alpha|v_1| + (1-\alpha)|v_1|$$
$$\leqq |v_1| \qquad (2.1.24)$$

and hence X is convex.

For $v \in X$,

$$|(Tv)(t)| \leqq |\sigma_2(t)| + \int_t^\infty v^2\,ds$$
$$\leqq |\sigma_1(t)| + \int_t^\infty v_1^2\,ds$$
$$= v_1(t) \qquad t \geqq t_0 . \qquad (2.1.25)$$

(2.1.25) implies that $TX \subset X$.

We now show that T is a continuous map from X into X . Let $(x_n) \subset X$ be such that $x_n \to x$ where $x \in X$. Since $|x_n| \leq v_1$ and $x \in L^2$

$$\left|\int_{t_0}^\infty (x_n^2 - x^2)\,ds\right| \leqq \int_{t_0}^\infty |x_n - x||x_n + x|\,ds$$
$$\leqq \|x_n - x\|\|x_n + x\|$$

by the Schwarz inequality.

$$\leqq \|x_n - x\|(\|x_n\| + \|x\|)$$

by Minkowski's inequality.

$$\leqq \|x_n - x\| (\|v_1\| + \|x\|) \; . \qquad (2.1.26)$$

Letting now $n \to \infty$ in (2.1.26) we see that

$$\int_{t_0}^{\infty} x_n^2 \, ds \to \int_{t_0}^{\infty} x^2 \, ds \; . \qquad (2.1.27)$$

The same argument shows that

$$\int_{t}^{\infty} x_n^2 \, ds \to \int_{t}^{\infty} x^2 \, ds \qquad (2.1.28)$$

for each $t \geqq t_0$.

Hence

$$\begin{aligned} (Tx_n)(t) &= \sigma_2(t) + \int_t^{\infty} x_n^2 \, ds \\ &\to \sigma_2(t) + \int_t^{\infty} x^2 \, ds \\ &\overset{df}{=} (Tx)(t) \qquad t \geqq t_0 \; . \end{aligned} \qquad (2.1.29)$$

This implies that $|Tx_n - Tx|^2 \to 0$, each $t \geqq t_0$, as $n \to \infty$. Moreover

$$|(Tx_n)(t) - (Tx)(t)|^2 \leqq 4v_1^2 \qquad (2.1.30)$$

whenever $t \geqq t_0$. Thus the Lebesgue dominated convergence theorem [24, p. 110] implies that

$$\int_t^{\infty} |Tx_n - Tx|^2 \to 0 \quad \text{as} \quad n \to \infty \tag{2.1.31}$$

for $t \geqq t_0$. Hence T is continuous.

To show that TX is compact we use Corollary II.1.2 of Appendix II. (II.1.4) is satisfied since if $x \in X$, $|Tx| \le v_1$ and so

$$\int_{t_0}^{\infty} |Tx|^2 \le \int_{t_0}^{\infty} v_1^2 = \|v_1\|^2 . \tag{2.1.32}$$

If we let $E_A = \{t : t_0 \le A \le t < \infty\}$ then given $\varepsilon > 0$, we choose A sufficiently large so that

$$\int_A^{\infty} v_1^2 < \varepsilon . \tag{2.1.33}$$

This will then imply that

$$\int_A^{\infty} |Tx|^2 < \varepsilon \tag{2.1.34}$$

for all $x \in X$ by virtue of (2.1.32). This proves (II.1.5). To prove (II.1.6-7) we need some additional information. Since v_1 is a solution of (2.1.20),

$$\begin{aligned} v_1^2(t) &= \left\{|\sigma_1(t)| + \int_t^{\infty} v_1^2\right\}^2 \\ &\geqq |\sigma_1(t)|^2 \qquad\qquad t \geqq t_0 \end{aligned} \tag{2.1.35}$$

and so $\sigma_1 \in L^2(t_0, \infty)$.

By the same argument,

$$v_1^2(t) \geqq \left\{\int_t^\infty v_1^2\right\}^2 \qquad t \geqq t_0$$

and so

$$\int_t^\infty v_1^2 \in L^2(t_0, \infty) \tag{2.1.36}$$

The following theorem [24, p.] will also be useful.

If $f \in L^p[t_0, \infty)$, $p \geqq 1$, then

$$\|f(x+h) - f(x)\|_p \to 0 \quad \text{as} \quad h \to 0 . \tag{2.1.37}$$

Since $\sigma_1 \in L^2[t_0, \infty)$ we have from (2.1.19) that $\sigma_2 \in L^2[t_0, \infty)$ and thus

$$\|\sigma_2(t+h) - \sigma_2(t)\| \to 0 \quad \text{as} \quad h \to 0 \tag{2.1.38}$$

on account of (2.1.37).

Similarly if we set

$$V(t) = \int_t^\infty v_1^2$$

then

$$\|V(t+h) - V(t)\| \to 0 \quad \text{as} \quad h \to 0 \tag{2.1.39}$$

because of (2.1.36).

Thus if $x \in X$, $\varepsilon > 0$

$$\int_{t_0}^{t_0+h} |Tx|^2 \leqq \int_{t_0}^{t_0+h} v_1^2 < \varepsilon \tag{2.1.40}$$

if $|h| < \delta$, by the continuity of the integral. This proves (II.1.6).

For $x \in X$, $\varepsilon > 0$

$$\begin{aligned} \|(Tx)(t+h) - (Tx)(t)\| &= \|\sigma_2(t+h) - \sigma_2(t) + \int_{t+h}^{t} x^2\,ds\| \\ &\leqq \|\sigma_2(t+h) - \sigma_2(t)\| + \|\int_{t+h}^{t} x^2\| . \end{aligned} \tag{2.1.41}$$

From (2.1.38) we can choose h so that if $|h| < \delta_1$ then

$$\|\sigma_2(t+h) - \sigma_2(t)\| < \frac{\varepsilon}{2} . \tag{2.1.42}$$

Similarly there is a $\delta_2 > 0$ such that whenever $|h| < \delta_2$

$$\|V(t+h) - V(t)\| < \frac{\varepsilon}{2} . \tag{2.1.43}$$

Thus

$$\begin{aligned} \|\int_{t+h}^{t} x^2\| &\leqq \|\int_{t+h}^{t} v_1^2\| \\ &= \|V(t+h) - V(t)\| \\ &< \frac{\varepsilon}{2} . \end{aligned} \tag{2.1.44}$$

Hence if $|h| < \delta = \min\{\delta_1, \delta_2\}$ then, for any $x \in X$,

$$\|(Tx)(t+h) - (Tx)(t)\| < \varepsilon$$

which proves (II.1.7) and therefore TX is compact. Consequently the Schauder theorem implies the existence of a fixed point $v = Tv$ for (2.1.21) and this completes the proof.

COROLLARY 2.1.2:

Let $\sigma_1(t) \geqq 0$ for $t \geqq t_0$ and

$$\sigma_1(t) \geqq |\sigma_2(t)| \qquad t \geqq t_0 . \tag{2.1.45}$$

If

$$v_1(t) = \sigma_1(t) + \int_t^\infty v_1^2\, ds \tag{2.1.46}$$

has a solution for $t \geqq t_0$ then

$$v(t) = \pm\sigma_2(t) + \int_t^\infty v^2\, ds \tag{2.1.47}$$

has a solution for $t \geqq t_0$.

Proof: This follows immediately from the theorem.

THEOREM 2.1.3:

With σ_1, σ_2 as above and

$$\sigma_1(t) \geqq |\sigma_2(t)| \qquad t \geqq t_0 \; , \qquad (2.1.48)$$

suppose that

$$y'(t) = c_1 + \int_a^t y(s)d\sigma_1(s) \qquad (2.1.49)$$

is non-oscillatory. Then

$$z'(t) = c_2 \pm \int_a^t z(s)d\sigma_2(s) \qquad (2.1.50)$$

is non-oscillatory.

Proof: This is immediate from Corollary 2.1.2 and Theorem 2.1.1.

THEOREM 2.1.4:

Let $\sigma(t)$ satisfy the conditions of Theorem 2.1.1. If

$$t|\sigma(t)| \leqq \frac{1}{4} \qquad t \geqq t_0 > 0 \qquad (2.1.51)$$

then (2.1.0) is non-oscillatory.

Proof: Let $\sigma_1(t) = 1/4\,t$ and $\sigma_2(t) \equiv \sigma(t)$ and apply Theorem 2.1.3. This is permissible since (2.1.49) is then equivalent to

$$y'' + \frac{1}{4t^2}\, y = 0 \qquad (2.1.52)$$

which is a non-oscillatory Euler equation [59, p. 45]. The result now follows.

COROLLARY 2.1.3:

(2.1.0) is non-oscillatory if

$$\limsup_{t\to+\infty} t|\sigma(t)| < \frac{1}{4} . \qquad (2.1.53)$$

Proof: This is immediate since (2.1.53) implies (2.1.51) if t is sufficiently large.

THEOREM 2.1.5:

Let σ_1, σ_2 be as in Theorem 2.1.2 and

$$\sigma_1(t) \geqq |\sigma_2(t)| \qquad t \geqq t_0 . \qquad (2.1.54)$$

If (2.1.49) is non-oscillatory then to every solution $y(t)$ of (2.1.49) there corresponds a solution $z(t)$ of (2.1.50) such that

$$z(t) \leqq |y(t)| \qquad t \geqq t^* . \qquad (2.1.55)$$

Proof: We first note that (2.1.50) is non-oscillatory on account of Theorem 2.1.3.

If $y(t)$ is a non-oscillatory solution of (2.1.49) then either $y(t) > 0$ or $y(t) < 0$ for $t \geqq t_1$. If $y(t) > 0$,

Theorem 2.1.1 implies that (2.1.46) has a solution $v_1(t)$ for $t \geqq t^* = \max\{t_0, t_1\}$. Hence (2.1.47) has a solution $v(t)$ for $t \geqq t^*$ (because of Corollary 2.1.2) which corresponds to some non-oscillatory solution $z(t)$ of (2.1.50). We can suppose $z(t) > 0$ for $t \geqq t^*$. Since the proof of Theorem 2.1.2 guarantees that

$$|v(t)| \leqq v_1(t) \qquad t \geqq t^* \tag{2.1.56}$$

we can recover the non-oscillatory solutions y, z to find that

$$z(t) \leqq y(t) \qquad t \geqq t^* . \tag{2.1.57}$$

If $z(t) < 0$ for $t \geqq t^*$ the last line is clear. On the other hand if $y(t) < 0$ for $t \geqq t_1$ then $-y(t) > 0$ and the above argument shows that there is some solution $z(t)$ such that

$$z(t) \leqq -y(t) \qquad t \geqq t^* . \tag{2.1.58}$$

This completes the proof.

<u>THEOREM 2.1.6</u>:

Let $\sigma(t)$ satisfy the hypotheses of Theorem 2.1.1. If

$$\int_t^\infty \sigma^2(s)ds \leqq \frac{1}{4}|\sigma(t)| \qquad t \geqq t_0 , \tag{2.1.59}$$

then (2.1.0) is non-oscillatory.

<u>Proof</u>: By Theorem 2.1.1 it suffices to show that (2.1.4) has a solution for sufficiently large t. We shall again make use of the Schauder fixed point theorem.

Let X be a subset of $L^2(t_0, \infty)$ defined by

$$X = \{v \in L^2(t_0, \infty) : |v(t) - \sigma(t)| \leqq |\sigma(t)|, t \geqq t_0\} . \tag{2.1.60}$$

For $v \in X$ we define a map T by

$$(Tv)(t) = \sigma(t) + \int_t^\infty v^2 \, ds . \tag{2.1.61}$$

If $\alpha \in [0, 1]$ and $u, v \in X$,

$$\begin{aligned} |\alpha u + (1-\alpha)v - \sigma| &= |\alpha(u-\sigma) + (1-\alpha)(v-\sigma)| \\ &\leqq \alpha|u-\sigma| + (1-\alpha)|v-\sigma| \\ &\leqq \alpha|\sigma| + (1-\alpha)|\sigma| \\ &\leqq |\sigma| . \end{aligned}$$

This shows that X is convex.

Moreover if $v \in X$ then

$$|v(t)| \leqq 2|\sigma(t)| \qquad t \geqq t_0 . \tag{2.1.62}$$

Hence

$$\begin{aligned} |(Tv)(t) - \sigma(t)| &= \int_t^\infty v^2\,ds \\ &\leqq 4 \int_t^\infty \sigma^2\,ds \qquad t \geqq t_0\ , \\ &\leqq 4 \cdot \frac{1}{4}\,|\sigma(t)| \qquad t \geqq t_0\ , \\ &< |\sigma(t)| \qquad t \geqq t_0 \end{aligned} \tag{2.1.63}$$

which implies that $TX \subset X$. The continuity of T is shown in exactly the same way as in Theorem 2.1.2 wherein we now make use of (2.1.62) instead of $|v(t)| \leqq v_1(t)$. The compactness of TX can be shown by applying Corollary II.1.2 to the set TX by making extensive use of (2.1.62). The procedure is similar to that in Theorem 2.1.2 and is therefore omitted. Consequently the Schauder theorem implies the existence of a fixed point $v = Tv$ of (2.1.61) and this completes the proof.

REMARK:

We note that σ can be replaced by $-\sigma$ in (2.1.49-50) and the conclusion will be the same with the appro priate change in (2.1.0).

THEOREM 2.1.7:

Let $\sigma(t)$ satisfy the hypotheses of Theorem 2.1.1 and

(2.1.59).

If $\sigma(t) \geqq 0$ then

$$z'(t) = c_2 + \int_a^t z(s)d\sigma_2(s) \qquad (2.1.64)$$

will be non-oscillatory where

$$\sigma_2(t) = 4 \int_t^\infty \sigma^2(s)ds . \qquad (2.1.65)$$

Proof: Let $\sigma_1(t) \equiv \sigma(t)$. Since $\sigma_2(t) \geqq 0$, (2.1.59) implies that

$$\sigma_1(t) \geqq \sigma_2(t) \qquad t \geqq t_0 . \qquad (2.1.66)$$

Therefore Theorem 2.1.6 shows that (2.1.0) is non-oscillatory. Theorem 2.1.3 now implies that (2.1.64) is non-oscillatory and this completes the proof.

COROLLARY 2.1.4:

Let $\sigma(t) \geqq 0$ satisfy the hypotheses of Theorem 2.1.7. Then (2.1.0) has a non-oscillatory solution $y(t)$ such that, for $t \geqq t_0$,

$$|y(t)| \leqq \exp\left\{ 2 \int_{t_0}^t \sigma(s)ds \right\} . \qquad (2.1.67)$$

Proof: The hypothesis implies that (2.1.0) is non-oscillatory.

The proof of Theorem 2.1.6 implies the existence of $v(t)$, a solution of (2.1.4), such that (2.1.62) holds. This estimate then implies (2.1.67) for y.

It is possible to remove the requirements that (2.1.59) hold and that $\sigma(t) \geqq 0$ in Theorem 2.1.7 and then state the converse.

<u>THEOREM 2.1.8</u>:

Let $\sigma(t)$ satisfy the hypotheses of Theorem 2.1.1. With σ_2 defined as in (2.1.65) suppose that (2.1.64) is non-oscillatory.

Then (2.1.0) is non-oscillatory and for each non-trivial solution z of (2.1.64) there is a solution y of (2.1.0) such that

$$0 < y(t) < |z(t)|^{\frac{1}{2}} \exp\left\{\int_{t_1}^{t} |\sigma(s)|ds\right\}$$

for $t \geqq t_1$ say.

<u>Proof</u>: We use Schauder's fixed point theorem. Consider the space $L^2(t_0, \infty)$ and a subset X defined by

$$X = \{v \in L^2(t_0, \infty) : |v(t)| \leqq \frac{1}{2} v_1(t) + |\sigma(t)| , t \geqq t_0\}$$

where $v_1(t)$ is a solution of the integral equation

$$v(t) = 4\int_t^\infty \sigma^2(s)ds + \int_t^\infty v^2(s)ds$$

which exists by virtue of Theorem 2.1.1. For $v \in X$ we define a map T by

$$(Tv)(t) = \sigma(t) + \int_t^\infty v^2 ds \qquad t \geqq t_0 .$$

As in Theorem 2.1.2 a simple calculation shows that X is convex. If $v \in X$, $t \geqq t_0$,

$$\begin{aligned} |(Tv)(t)| &\leqq |\sigma(t)| + \int_t^\infty v^2\, ds \\ &\leqq |\sigma(t)| + \int_t^\infty \{\tfrac{1}{2}v_1 + |\sigma|\}^2 ds \\ &\leqq |\sigma(t)| + 2\cdot\tfrac{1}{4}\int_t^\infty v_1^2 ds + 2\int_t^\infty \sigma^2 ds \\ &\leqq |\sigma(t)| + \tfrac{1}{2}v_1(t) . \end{aligned}$$

Hence $TX \subset X$. The continuity of T and the compactness of TX can be shown analogously as in Theorem 2.1.2 relying heavily upon the estimate

$$v \in X: \qquad |v(t)| \leqq \tfrac{1}{2}v_1(t) + |\sigma(t)| \qquad t \geqq t_0 .$$

Thus Schauder's theorem implies that T has a fixed point $v = Tv$ which necessarily satisfies the latter inequality. This means that (2.1.0) is non-oscillatory and we can recover an eventually positive solution $y(t)$ of (2.1.0) from the integral equation such that

$$0 < y(t) < |z(t)|^{\frac{1}{2}} \exp\left\{\int_{t_1}^{t} |\sigma(s)|ds\right\} .$$

2.1A APPLICATIONS TO DIFFERENTIAL EQUATIONS:

Let $a(t)$, $b(t)$ be continuous functions on $[a, \infty)$ and suppose that the integrals

$$A(t) = \int_t^\infty a(s)ds \tag{2.1.68}$$

$$B(t) = \int_t^\infty b(s)ds \tag{2.1.69}$$

exist, in the limiting sense.

Consider the equations

$$y'' + a(t)y = 0 \tag{2.1.70}$$

$$z'' + b(t)z = 0 . \tag{2.1.71}$$

The following results are all consequences of section 2.1.

THEOREM 2.1.1A:

Let $a(t)$ be as above. Then a necessary and sufficient condition for (2.1.70) to be non-oscillatory is that the non-linear integral equation

$$v(t) = \int_t^\infty a(s)ds + \int_t^\infty v^2(s)ds \qquad (2.1.72)$$

have a solution for sufficiently large t .

Proof: This follows immediately from Theorem 2.1.1 upon letting $\sigma(t) \equiv A(t)$ in (2.1.4) and noticing that (2.1.0) is then equivalent to (2.1.70).

This result extends the theorem of Hille [31, p. 243] where it was assumed that $a(t) > 0$. The existence of a function $v(t)$ satisfying (2.1.72) is reminiscent of Wintner's criterion [62, p. 375].

THEOREM 2.1.2A:

Let $a(t)$, $b(t)$ be defined as above and suppose that

$$|A(t)| \geqq |B(t)| \qquad t \geqq t_0 . \qquad (2.1.73)$$

If

$$v_1(t) = |A(t)| + \int_t^\infty v_1^2\,ds \qquad (2.1.74)$$

has a solution for $t \geqq t_0$ then

$$v(t) = B(t) + \int_t^\infty v^2\,ds \qquad (2.1.75)$$

also has a solution for $t \geqq t_0$.

Proof: This is immediate from Theorem 2.1.2 upon setting $\sigma_1(t) \equiv A(t)$, $\sigma_2(t) \equiv B(t)$.

COROLLARY 2.1.2A:

Let $A(t) \geqq 0$ for $t \geqq t_0$ and

$$A(t) \geqq |B(t)| \qquad t \geqq t_0 . \tag{2.1.76}$$

If

$$v_1(t) = A(t) + \int_t^{\infty} v_1^2\, ds \tag{2.1.77}$$

has a solution then (2.1.75) has a solution.

Proof: Immediate from the theorem.

When $a(t) \geqq 0$, $b(t) \geqq 0$ Theorem 2.1.2A was proven by Hille [31, p. 245].

THEOREM 2.1.3A: Let $a(t)$, $b(t)$ satisfy the usual conditions (those stated in the beginning of this subsection). If

$$A(t) \geqq |B(t)| \qquad t \geqq t_0 \tag{2.1.78}$$

and (2.1.70) is non-oscillatory then (2.1.71) is also non-oscillatory.

Proof: This is readily verified by setting $\sigma_1(t) \equiv A(t)$ and $\sigma_2(t) \equiv B(t)$ in Theorem 2.1.3.

As it stands, the previous theorem is due to Taam [60, p. 495]. (See also [25, p. 369, exercise 7.9].) In fact, Taam stated a slightly more general theorem than the above as he considered equations of the form (1.0.1). The case $A(t) \geqq B(t) \geqq 0$ is due to Wintner [63, p. 257] who extended Hille's [31, p. 245] criterion $a(t) \geqq 0$, $b(t) \geqq 0$ along with $A(t) \geqq B(t)$.

THEOREM 2.1.4A:

Let $a(t)$ satisfy the usual conditions. If

$$t\left|\int_t^\infty a(s)ds\right| \leqq \frac{1}{4} \qquad t \geqq t_0 > 0 , \qquad (2.1.79)$$

then (2.1.70) is non-oscillatory.

Proof: This follows immediately from Theorem 2.1.4 upon setting $\sigma(t) \equiv A(t)$.

COROLLARY 2.1.3A:

(2.1.70) is non-oscillatory if

$$\limsup_{t\to\infty} t\left|\int_t^\infty a(s)ds\right| < \frac{1}{4} . \qquad (2.1.80)$$

<u>Proof</u>: Set $\sigma(t) = A(t)$ in Corollary 2.1.3.

Wintner [62, p. 370] essentially extends theorem 2.1.4A by replacing (2.1.79) with

$$-\frac{3}{4} \leqq t \int_t^\infty a(s)ds \leqq \frac{1}{4} \qquad t \geqq t_0 . \qquad (2.1.81)$$

Thus he improved the lower bound of $-\frac{1}{4}$ appearing in (2.1.79) to $-\frac{3}{4}$. The question seems to be open as to whether the latter number is best possible or not. When $a(t) > 0$ Theorem 2.1.4A was proven by Hille [31, p. 246]. For $a(t) > 0$, Corollary 2.1.3A can be found in [31, p. 246] where it is also shown that the bound appearing in (2.1.80) is best possible [31, pp. 248-49].

<u>THEOREM 2.1.5A</u>:

Let $a(t)$, $b(t)$ be continuous on $[a, \infty)$ and suppose that the integrals (2.1.68-69) are convergent (possibly conditionally). Suppose further that

$$A(t) \geqq |B(t)| \qquad t \geqq t_0 . \qquad (2.1.82)$$

If (2.1.70) is non-oscillatory then to every solution $y(t)$ of (2.1.70) there corresponds a solution $z(t)$ of (2.1.71) such that

$$z(t) \leqq |y(t)| \qquad t \geqq t^* . \qquad (2.1.83)$$

Proof: This follows from Theorem 2.1.5 via the substitutions in the proof of Theorem 2.1.3A.

The conclusion of the above theorem also holds if we only assume $a(t) \geqq b(t)$ where $a(t)$ is unrestricted to sign. The latter is due to Hartman and Wintner [26, p. 635]. Note that (2.1.82) requires $A(t) \geqq 0$ for large t but this need not be so for $a(t)$. Thus estimates for solutions of (2.1.71) can be obtained from (2.1.83), under the above hypotheses, whenever the solutions of (2.1.70) are known or can be estimated. (See for example [26, p. 636]).

THEOREM 2.1.6A:

Let $a(t)$ be continuous on $[a, \infty)$ and suppose that the integral $A(t)$ converges (possibly conditionally). If for $t \geqq t_0$

$$\int_t^\infty A^2(s)ds \leqq \frac{1}{4}|A(t)| , \qquad (2.1.84)$$

then (2.1.70) is non-oscillatory.

Proof: This follows immediately from Theorem 2.1.6 upon setting $\sigma(t) \equiv A(t)$.

When $A(t) \geqq 0$ the above theorem was proven by Opial [47, p. 312] and extended a result of Wintner [62, p. 371] which states that (2.1.70) is non-oscillatory if

$$A^2(t) \leqq \frac{1}{4} a(t) \qquad t \geqq t_0 .$$

Thus in Theorem 2.1.6A $A(t)$ is no longer required to be non-negative. Equality in (2.1.84) is attained in the case of the Euler equation (2.1.52).

THEOREM 2.1.7A:

Let $a(t)$ satisfy the hypotheses of Theorem 2.1.6A along with (2.1.84). If $A(t) \geqq 0$ for large t then

$$y'' + 4A^2(t)y = 0 \tag{2.1.85}$$

is non-oscillatory.

Proof: Refer to Theorem 2.1.7 with $\sigma(t) \equiv A(t)$.

Whether (2.1.68) being non-oscillatory implies that (2.1.85) is, appears to be an open question [59, p. 93] which we shall discuss in section 2.2.

COROLLARY 2.1.4A:

Let $A(t) \geqq 0$ and suppose that (2.1.84) is satisfied. Then (2.1.68) has a non-oscillatory solution $y(t)$ such that, for $t \geqq t_0$,

$$|y(t)| \leqq \exp\left\{2\int_{t_0}^{t} A(s)ds\right\} . \tag{2.1.86}$$

Proof: This follows from Corollary 2.1.4 with $\sigma(t) \equiv A(t)$.

THEOREM 2.1.8A:

Let $A(t)$ be defined as in (2.1.68) and suppose that

$$z'' + 4A^2(t)z = 0 \qquad (2.1.87)$$

is non-oscillatory. Then

$$y'' + a(t)y = 0 \qquad (2.1.88)$$

is non-oscillatory and for each non-trivial solution $z(t)$ of (2.1.87) there is a solution $y(t)$ of (2.1.88) such that

$$0 < y(t) < |z(t)|^{\frac{1}{2}} \exp\left\{\int_{t_1}^{t} |A(s)|ds\right\} \qquad (2.1.89)$$

for t sufficiently large, say, $t \geq t_1$.

Proof: This is an application of Theorem 2.1.8.

The first part of the theorem is identical with a theorem of Hartman and Wintner [27, p. 216] though the estimate (2.1.89) is stronger than the corresponding estimate in [27] where the absolute value sign about $A(t)$ in (2.1.89) does not appear. Thus the first part of Theorem 2.1.8 extends the Hartman-Wintner result cited above to equations of the type (2.1.0), while the second part extends the corresponding result only when $\sigma(t) \geqq 0$ in Theorem 2.1.8.

2.1B APPLICATIONS TO DIFFERENCE EQUATIONS:

In this subsection we apply the theorems of section 2.1 to recurrence relations of the form

$$c_n y_{n+1} + c_{n-1} y_{n-1} + b_n y_n = 0 \tag{2.1.90}$$

where $c_n > 0$, $n = -1, 0, 1, \ldots$, (b_n) is any given real sequence, $n = 0, 1, \ldots$.

We shall assume, unless otherwise specified, that

$$\sum_{0}^{\infty} \frac{1}{c_{n-1}} = \infty \tag{2.1.91}$$

be satisfied as an extra condition upon the c_n.

We saw in Chapter 1 that if $\sigma(t)$ is a step-function with jumps, at a fixed increasing sequence of points (t_n) where $t_{-1} = a$ and

$$t_n - t_{n-1} = \frac{1}{c_{n-1}} \qquad n = 0, 1, \ldots, \tag{2.1.92}$$

of magnitude

$$\sigma(t_n) - \sigma(t_n - 0) = -b_n - c_n - c_{n-1} \tag{2.1.93}$$

for $n = 0, 1, 2, \ldots$, then (2.1.0) gives rise to solutions of some "extended" recurrence relation in the sense that the resulting solution is a polygonal curve $y(t)$ defined on $[a, \infty)$ which has the property that if we write $y_n \equiv y(t_n)$, then the sequence (y_n) is a solution to the three-term recurrence relation (2.1.90) for $n = 0, 1, \ldots$.

We note that, whenever $\sigma(t)$ is defined by (2.1.93) for given sequences c_n, b_n, then for $t \in [t_m, t_{m+1})$, $m \geqq 0$,

$$\sigma(t) = \sigma(a) - \sum_{0}^{m} (b_n + c_n + c_{n-1}) . \qquad (2.1.94)$$

This follows from (2.1.93) and the relation $\sigma(t_n - 0) = \sigma(t_{n-1})$.

THEOREM 2.1.1B:

Let $\sigma(t)$ be defined as in (2.1.94) and assume that $\sigma(\infty)$ exists and is zero. Then a necessary and sufficient condition for (2.1.90) to be non-oscillatory is that

$$v(t) = \sigma(t) + \int_t^{\infty} v^2\, ds \qquad (2.1.95)$$

where $\sigma(t)$ is given by (2.1.94), have a solution which is in L^2 at infinity.

Proof: This follows immediately from Theorem 2.1.1 and the results of Chapter 1.

REMARK:

A solution (y_n) of (2.1.90) is said to be oscillatory

if the sequence exhibits an infinite number of sign changes and non-oscillatory if, for $n \geq N$, the sequence retains a constant sign. The discrete version of the Sturm separation theorem shows that if a solution is oscillatory (non-oscillatory) then all solutions inherit the same property. Moreover the transition from (2.1.0) to (2.1.90), in the case when σ is given by (2.1.94), shows that a given solution of (2.1.0) is oscillatory (non-oscillatory) if and only if the corresponding solution of (2.1.90) is oscillatory (non-oscillatory).

Thus, with σ defined as in (2.1.94),

$$y'(t) = c + \int_a^t y(s)d\sigma(s) \qquad t \in [a, \infty), \tag{2.1.96}$$

is oscillatory (non-oscillatory) if and only if

$$c_n y_{n+1} + c_{n-1}y_{n-1} + b_n y_n = 0 \qquad n = 0, 1, \ldots \tag{2.1.97}$$

is oscillatory (non-oscillatory).

The latter theorem thus gives the discrete version of Hille's theorem [31].

For given sequences $c_n > 0$, b_n, g_n we define step functions σ_1, σ_2 on $[a, \infty)$ by setting

$$\sigma_1(t) = \sigma_1(a) - \sum_0^m (b_n + c_n + c_{n-1}) \tag{2.1.98}$$

if $t \in [t_m , t_{m+1})$, $m \geqq 0$, and

$$\sigma_2(t) = \sigma_2(a) - \sum_0^m (g_n + c_n + c_{n-1}) \qquad (2.1.99)$$

if $t \in [t_m , t_{m+1})$. We recall that the c_n also satisfy (2.1.92) a fortiori, so that $t_n \to \infty$ as $n \to \infty$.

With σ_1 , σ_2 so defined we obtain the discrete analogs of Theorem 2.1.2 and Corollary 2.1.2 denoted by Theorem 2.1.2B and Corollary 2.1.2B respectively. Since the latter two results can be stated in the same way as the former two, we shall omit them and it shall be understood that when we refer to either of Theorem 2.1.2B or its corollary we shall mean Theorem 2.1.2 or its corollary with σ_1 , σ_2 given by (2.1.98-99).

THEOREM 2.1.3B:

Let $c_n > 0$ and satisfy (2.1.91). Suppose that

$$\lim_{m\to\infty} \sum_0^m (b_n + c_n + c_{n-1}) \qquad (2.1.100)$$

$$\lim_{m\to\infty} \sum_0^m (g_n + c_n + c_{n-1}) \qquad (2.1.101)$$

both exist and are finite (so that the series need only be conditionally convergent).

Suppose further that

$$\sum_{m}^{\infty} (c_n + c_{n-1} + b_n) \geqq \left| \sum_{m}^{\infty} (c_n + c_{n-1} + g_n) \right| \qquad (2.1.102)$$

for $m \geqq m_0$. If

$$c_n y_{n+1} + c_{n-1} y_{n-1} + b_n y_n = 0 \qquad (2.1.103)$$

is non-oscillatory then

$$c_n z_{n+1} + c_{n-1} z_{n-1} + g_n z_n = 0 \qquad (2.1.104)$$

is non-oscillatory.

<u>Proof</u>: Define σ_1 , σ_2 by (2.1.98), (2.1.99) respectively. (2.1.100-101) are then equivalent to requiring that both $\sigma_1(\infty)$, $\sigma_2(\infty)$ exist and be finite. Since we can alter these by an additive factor, we can assume that $\sigma_1(\infty) = \sigma_2(\infty) = 0$. This then implies that

$$\sigma_1(a) = \sum_{0}^{\infty} (c_n + c_{n-1} + b_n) \qquad (2.1.105)$$

$$\sigma_2(a) = \sum_{0}^{\infty} (c_n + c_{n-1} + g_n) \ . \qquad (2.1.106)$$

Hence, for $t \in [t_{m-1}, t_m)$,

$$\sigma_1(t) = \sum_{m}^{\infty} (c_n + c_{n-1} + b_n) \qquad (2.1.107)$$

$$\sigma_2(t) = \sum_{m}^{\infty} (c_n + c_{n-1} + g_n) \qquad (2.1.108)$$

Thus the requirement that (2.1.48) be satisfied for large t is equivalent to the requirement that (2.1.102) hold for large m. From the remark we see that (2.1.49) must be non-oscillatory. Hence Theorem 2.1.3 applies and hence (2.1.50) is non-oscillatory. Consequently (2.1.104) is also non-oscillatory and this completes the proof.

The latter theorem is therefore the discrete analog of the Taam result [60]. Simultaneously it provides an extension of the discrete version of the theorem of Wintner [63] and Hille [31] (see Theorem 2.1.3A). Thus for example, if

$$b_n \geqq 0 \qquad n = 0, 1, \ldots \tag{2.1.109}$$

$$g_n \geqq 0 \qquad n = 0, 1, \ldots \tag{2.1.110}$$

and

$$\sum_m^\infty b_n \geqq \sum_m^\infty g_n \qquad m \geqq m_0 , \tag{2.1.111}$$

then (2.1.104) is non-oscillatory if (2.1.103) is. This would be the formulation of the discrete analog of Hille's theorem [31].

THEOREM 2.1.4B:

Let $c_n > 0$ and satisfy (2.1.91). For a given sequence (b_n) assume that (2.1.100) exists. If

$$\left\{\sum_0^m \frac{1}{c_{n-1}}\right\}\left|\sum_m^\infty (c_n + c_{n-1} + b_n)\right| \leqq \frac{1}{4} \qquad m \geqq m_0 \qquad (2.1.112)$$

then (2.1.90) is non-oscillatory.

<u>Proof</u>: We define σ by (2.1.93). Then, for $t \in [t_{m-1}, t_m)$, we shall have

$$\sigma(t) = \sum_m^\infty (c_n + c_{n-1} + b_n) \ . \qquad (2.1.113)$$

For (2.1.51) to hold for large t it is necessary that

$$t\left|\sum_m^\infty (c_n + c_{n-1} + b_n)\right| \leqq \frac{1}{4} \qquad (2.1.114)$$

for all $t \in [t_{m-1}, t_m)$ when m is sufficiently large. Thus we let $t \to t_m$ in (2.1.114) and use (2.1.92) to obtain

$$\left\{a + \sum_0^m \frac{1}{c_{n-1}}\right\}\left|\sum_m^\infty (c_n + c_{n-1} + b_n)\right| \leqq \frac{1}{4} \qquad (2.1.115)$$

for $m \geqq m_0$. Without loss of generality we can assume that $t_{-1} = a = 0$. Then (2.1.115) will imply that (2.1.51) is satisfied for large t and hence that (2.1.96) is non-oscillatory by Theorem 2.1.4. Consequently (2.1.90) is non-oscillatory which is what we wished to prove.

In the particular case when $c_n = 1$ for all n and $t_{-1} = -1$ we obtain $t_n = n$ for all $n \geqq 0$. By replacing b_n in (2.1.90) by $b_n - 2$ we obtain a non-oscillation criterion for the equation

$$\Delta^2 y_{n-1} + b_n y_n = 0 \qquad n = 0 , 1 , \ldots . \qquad (2.1.116)$$

The latter theorem shows that (2.1.116) is non-oscillatory if

$$m\left|\sum_m^\infty b_n\right| \leqq \frac{1}{4} \qquad m \geqq m_0 . \qquad (2.1.117)$$

<u>Example 1</u>: Let $b_n = \gamma(n+1)^{-2}$ $n = 0 , 1 , \ldots$ in (2.1.116). Then

$$\Delta^2 y_{n-1} + \frac{\gamma}{(n+1)^2} y_n = 0 \qquad (2.1.118)$$

is non-oscillatory if $\gamma \leqq \frac{1}{4}$.

For

$$m\left|\sum_m^\infty b_n\right| = m \sum_m^\infty \frac{\gamma}{(n+1)^2}$$

$$\leqq m\gamma \cdot \int_m^\infty x^{-2}\, dx$$

$$\leqq \frac{m\gamma}{m+1}$$

$$\leqq \frac{1}{4} \qquad m \geqq 0 .$$

Consequently (2.1.117) holds with $m_0 = 0$ and thus (2.1.118) is non-oscillatory where $\gamma \leqq \frac{1}{4}$. This is the discrete Euler equation.

Example 2: Let $b_n = \gamma(-1)^n/(n+1)$, $n = 0, 1, \ldots$.

Then

$$\Delta^2 y_{n-1} + \frac{\gamma(-1)^n}{(n+1)} y_n = 0 \qquad (2.1.119)$$

is non-oscillatory if $|\gamma| \leqq \frac{1}{4}$.

For $\sum^{\infty} b_n$ is conditionally convergent and

$$m\left|\sum_m^\infty b_n\right| = m|\gamma|\left|\sum_m^\infty \frac{(-1)^n}{(n+1)}\right|$$

$$\leqq |\gamma| \left\{\frac{1}{(m+1)} - \left(\frac{1}{m+2} - \frac{1}{m+3} + \cdots\right)\right\}$$

$$\leqq |\gamma| \frac{m}{m+1}$$

$$\leqq \frac{1}{4} \qquad \text{if} \quad m \geqq 0 .$$

Consequently (2.1.117) applies and so (2.1.119) is non-oscillatory.

COROLLARY 2.1.3B:

If

$$\limsup_{m\to\infty} \left\{\sum_0^m \frac{1}{c_{n-1}}\right\}\left|\sum_m^\infty (c_n + c_{n-1} + b_n)\right| < \frac{1}{4} \qquad (2.1.120)$$

then (2.1.90) is non-oscillatory.

Proof: Follows from Corollary 2.1.3. i.e. (2.1.120) implies

that (2.1.112) holds for large m . In particular if

$$\limsup_{m\to\infty} m\left|\sum_{m}^{\infty} b_n\right| < \frac{1}{4} \qquad (2.1.121)$$

then (2.1.116) is non-oscillatory (because of (2.1.117)). (2.1.121) extends a result of Hinton and Lewis [32, p. 427] in which the series for b_n is required to be absolutely convergent. Because of the results in the next section it would appear that the upper bound $\frac{1}{4}$ in (2.1.121) is best possible, i.e. there exists (b_n) such that equality holds in (2.1.121) but (2.1.116) is oscillatory.

<u>THEOREM 2.1.5B</u>:

Let the sequences c_n , b_n satisfy the hypotheses of Theorem 2.1.3B along with (2.1.100-101-102). If (2.1.103) is non-oscillatory then to every solution y_n of (2.1.103) there corresponds a solution z_n of (2.1.104) such that

$$z_n \leqq |y_n| \qquad n \geqq N . \qquad (2.1.122)$$

<u>Proof</u>: Define $\sigma_1(t)$, $\sigma_2(t)$ as in the proof of Theorem 2.1.3B. The result then follows from Theorem 2.1.5 where we set $z(t_n) \equiv z_n$, similarly for y , and use (2.1.55) to obtain (2.1.122).

<u>THEOREM 2.1.6B</u>:

Let the sequences c_n , b_n satisfy the hypotheses of

Theorem 2.1.4B.

If, for $m \geqq m_0$,

$$\sum_{i=m}^{\infty} \frac{1}{c_i} \left\{ \sum_{j=i+1}^{\infty} (c_j + c_{j-1} + b_j) \right\}^2 \leqq \frac{1}{4} \left| \sum_{i=m+1}^{\infty} (c_i + c_{i-1} + b_i) \right| \tag{2.1.123}$$

then (2.1.103) is non-oscillatory.

Proof: We define σ as in the proof of Theorem 2.1.4B. Then for $t \in [t_{m-1}, t_m)$, σ is given by (2.1.113). Consequently

$$\int_t^{\infty} \sigma^2(s)ds = \sum_{i=m-1}^{\infty} \int_{t_i}^{t_{i+1}} \sigma^2(s)ds - \int_{t_{m-1}}^{t} \sigma^2(s)ds . \tag{2.1.124}$$

Since σ is constant on each $[t_{m-1}, t_m)$, $m = 0, 1, \ldots$ we obtain

$$\int_t^{\infty} \sigma^2(s)ds = \sum_{i=m-1}^{\infty} (t_{i+1} - t_i) \left\{ \sum_{j=i+1}^{\infty} (c_j + c_{j-1} + b_j) \right\}^2$$

$$-(t - t_{m-1}) \left\{ \sum_{j=m}^{\infty} (c_j + c_{j-1} + b_j) \right\}^2 \tag{2.1.125}$$

$$= \sum_{i=m-1}^{\infty} \frac{1}{c_i} G_{i+1}^2 - (t - t_{m-1}) G_m^2 \tag{2.1.126}$$

where

$$G_i = \sum_{j=i}^{\infty} (c_j + c_{j-1} + b_j) .$$

Since $t \in [t_{m-1}, t_m)$ the second term in (2.1.126) can be neglected and so

$$\int_t^\infty \sigma^2(s)ds \leqq \sum_{i=m-1}^{\infty} \frac{1}{c_i} G_{i+1}^2 \quad , \quad t \in [t_{m-1}, t_m) \quad (2.1.127)$$

$$\leqq \frac{1}{4} |G_m| = \frac{1}{4} |\sigma(t)| \quad (2.1.128)$$

since $t \in [t_{m-1}, t_m)$. Thus (2.1.59) is satisfied for large t and so (2.1.0) is non-oscillatory. Consequently (2.1.103) is non-oscillatory.

From the discussion following the proof of Theorem 2.1.4B we can obtain a non-oscillation criterion for (2.1.116). We therefore find that if

$$\sum_{i=m}^{\infty} \left\{\sum_{j=i+1}^{\infty} b_j\right\}^2 \leqq \frac{1}{4}\left|\sum_{j=m+1}^{\infty} b_j\right| \quad (2.1.129)$$

for $m \geqq m_0$, then (2.1.116) is non-oscillatory. This follows from the above theorem. The numerical bound in (2.1.129) is best possible. To see this let the sequence b_n be defined as

$$b_n = \begin{cases} 0 & n = 0 \\ \frac{1}{4} & n = 1 \ . \\ 0 & n \geqq 2 \end{cases} \quad (2.1.130)$$

We then notice that we have equality in (2.1.129) when $m = 0$.

A simple computation now shows that the solution of (2.1.116) corresponding to the initial values $y_{-1} = 0$, $y_0 = 1$ admits the lower bound

$$y_n \geqq n-1 \qquad n \geqq 1 .$$

Hence (2.1.116) is non-oscillatory.

THEOREM 2.1.7B: With the c_n , b_n as in Theorem 2.1.4B assume that (2.1.123) is satisfied for large m . If

$$\sigma(t) \equiv \sum_{i=m}^{\infty} (c_i + c_{i-1} + b_i) \geqq 0 \quad , \quad t \in [t_{m-1}, t_m) \tag{2.1.131}$$

then the differential equation

$$z'' + 4\sigma^2(t) z = 0 \tag{2.1.132}$$

is non-oscillatory.

Proof: This follows immediately from Theorem 2.1.7.

COROLLARY 2.1.4B:

Let c_n , b_n satisfy the hypotheses of Theorem 2.1.7B and suppose that (2.1.131) holds. Then (2.1.90) has a non-oscillatory solution (y_n) such that for $n \geqq N$,

$$|y_n| \leqq \exp\left\{ 2 \int_T^{t_n} \sigma(s)\,ds \right\} \tag{2.1.133}$$

where σ is as in (2.1.131).

Proof: Follows from Corollary 2.1.4.

THEOREM 2.1.8B:

Let $\sigma(t)$ be as in Theorem 2.1.7B except that σ need not be non-negative. Assume that

$$z'' + 4\sigma^2(t)z = 0 \tag{2.1.134}$$

is non-oscillatory. Then

$$c_n y_{n+1} + c_{n-1}y_{n-1} + b_n y_n = 0 \tag{2.1.135}$$

is non-oscillatory and for each non-trivial solution z of (2.1.134) there is a solution (y_n) of (2.1.135) such that

$$0 < y_n < |z_n|^{\frac{1}{2}} \exp\left\{ \int_T^{t_n} |\sigma(s)| ds \right\}$$

where $z_n \equiv z(t_n)$ and σ is given by (2.1.131).

Proof: Follows from Theorem 2.1.8.

We can therefore summarize the results of Theorem 2.1.7B and 2.1.8B as follows:

For given sequences c_n, b_n satisfying (2.1.123), (2.1.135) will be non-oscillatory (Theorem 2.1.6B). If, in addition, we assume that (2.1.131) is satisfied then (2.1.132) will be non-oscillatory (Theorem 2.1.7B). Thus Theorem 2.1.7B gives a criterion which will ensure that a certain differential equation is non-oscillatory if a certain related recurrence relation is non-oscillatory. On the other hand, Theorem 2.1.8B gives the converse. That is, if a certain differential equation is non-oscillatory then a related recurrence relation will also be non-oscillatory.

§2.2 OSCILLATION CRITERIA:

In this section we continue the investigation of equations of the form (2.1.0) and complement some of the non-oscillation theorems in section 2.1 with oscillation theorems.

THEOREM 2.2.1:

Let σ satisfy the conditions of Theorem 2.1.1 and suppose that

$$\sigma(t) \geqq 0 \qquad t \geqq t_0 \,. \tag{2.2.1}$$

If

$$t\,\sigma(t) \leqq \frac{1}{4} \qquad t \geqq t_0 \tag{2.2.2}$$

then (2.1.0) is non-oscillatory.

If $\varepsilon > 0$ and

$$t\,\sigma(t) \geqq \frac{1}{4} + \varepsilon \qquad t \geqq t_0 \tag{2.2.3}$$

then (2.1.0) is oscillatory.

<u>Proof</u>: The first part is Theorem 2.1.4. To prove the second part we write $\sigma_1(t) \equiv \sigma(t)$ and

$$\sigma_2(t) \equiv \frac{1}{t}\left(\frac{1}{4}+\varepsilon\right) \qquad t \geqq t_0 . \tag{2.2.4}$$

(2.2.3) then implies that

$$\sigma_1(t) \geqq \sigma_2(t) \qquad t \geqq t_0 . \tag{2.2.5}$$

Furthermore

$$z'(t) = c + \int_a^t z(s)d\sigma_2(s) \tag{2.2.6}$$

is oscillatory since it is equivalent to

$$z'' + \frac{1}{t^2}\left(\frac{1}{4}+\varepsilon\right)z = 0 \tag{2.2.7}$$

and the latter equation is oscillatory for $\varepsilon > 0$ [59, p. 45]. Since (2.1.0) must either be oscillatory or non-oscillatory it cannot be non-oscillatory for then Theorem 2.1.3 would imply that (2.2.6) is non-oscillatory which is impossible. Thus (2.1.0) is oscillatory and the theorem is proved.

<u>THEOREM 2.2.2</u>:

Let σ satisfy the hypotheses of Theorem 2.1.1 and

assume that (2.2.1) holds.

If

$$\int_t^\infty \sigma^2(s)ds \leqq \frac{1}{4}\sigma(t) \qquad t \geqq t_0 \tag{2.2.8}$$

then (2.1.0) is non-oscillatory.

If $\varepsilon > 0$ and

$$\int_t^\infty \sigma^2(s)ds \geqq \left(\frac{1}{4}+\varepsilon\right)\sigma(t) \qquad t \geqq t_0\ , \tag{2.2.9}$$

then (2.1.0) is oscillatory.

<u>Proof</u>: The first part is Theorem 2.1.6. To prove that (2.1.0) is oscillatory under (2.2.9) we assume, on the contrary, that it is non-oscillatory. Then there exists a solution $y(t)$ of (2.1.0) such that

$$y(t) > 0 \qquad t \geqq T\ . \tag{2.2.10}$$

This then implies that the integral equation

$$v(t) = \sigma(t) + \int_t^\infty v^2\,ds \qquad t \geqq T \tag{2.2.11}$$

admits a solution $v(t)$ for $t \geqq T$. This solution $v(t)$ is non-negative for $t \geqq T$ and $v(t) \to 0$ as $t \to \infty$. Moreover

$$v(t) \geqq \sigma(t) \qquad t \geqq T\ . \tag{2.2.12}$$

Set $\alpha = (\frac{1}{4}) + \varepsilon$. (2.2.9) then implies that

$$\int_t^\infty \sigma^2(s)ds \geqq \alpha\sigma(t) \qquad t \geqq t_0 . \qquad (2.2.13)$$

Now,

$$v(t) \geqq \sigma(t) + \int_t^\infty \sigma^2(s)ds \qquad t \geqq T \qquad (2.2.14)$$

$$\geqq (1+\alpha)\sigma(t) \equiv \alpha_1\sigma(t) \qquad t \geqq T . \qquad (2.2.15)$$

We may take it that

$$\int_t^\infty \sigma^2(s)ds < \infty$$

since $v \in L^2$ and $v \geqq \sigma$ for large t .

Using (2.2.15) in (2.2.11) we obtain

$$v(t) \geqq \sigma(t) + \alpha_1^2 \int_t^\infty \sigma^2(s)ds \qquad (2.2.16)$$

$$\geqq \sigma(t) + \alpha_1^2\alpha\sigma(t)$$

$$\geqq (1+\alpha_1^2\alpha)\sigma(t)$$

$$\equiv \alpha_2\sigma(t) \qquad t \geqq T . \qquad (2.2.17)$$

Repeating the above process we find that

$$v(t) \geqq \alpha_n\sigma(t) \qquad t \geqq T \qquad (2.2.18)$$

where

$$\alpha_n = 1 + \alpha\alpha_{n-1}^2 \qquad n \geqq 2 , \tag{2.2.19}$$

is independent of t . A simple induction argument shows that (α_n) is increasing and hence tends to some limit β which must be infinity.

For if $\beta < \infty$ then, by (2.2.19),

$$\beta = 1 + \alpha\beta^2 \tag{2.2.20}$$

and since $\alpha > \frac{1}{4}$ no such β can exist. Hence $\beta = \infty$. Thus letting $n \to \infty$ in (2.2.18) we find that $v(t)$ must be unbounded at every point t where $\sigma(t) \neq 0$. This contradicts the boundedness of v on finite intervals. This contradiction shows that (2.1.0) cannot have a non-oscillatory solution and hence (2.1.0) must be oscillatory.

The previous theorems therefore give "ε-necessary and sufficient" conditions for (2.1.0) to be non-oscillatory. Thus because of Theorem 2.2.2 we see that the condition that (2.1.59) hold in Theorem 2.1.7 is not too restrictive. We can therefore complement Theorem 2.1.8 with the result stated in Theorem 2.1.7.

<u>THEOREM 2.2.3</u>:

Let σ be right-continuous and locally of bounded

variation on $[a, \infty)$.

If

$$\lim_{t\to\infty} \sigma(t) = -\infty \tag{2.2.21}$$

then (2.1.0) is oscillatory.

Proof: Suppose, on the contrary, that $y(t)$ is a non-oscillatory solution of (2.1.0) with say $y(t) > 0$ for $t \geqq t_0$. If $v(t) = y'(t)/y(t)$, $t \geqq t_0$, then from Theorem 2.1.1 we have

$$v(t) = \sigma(t) - \sigma(t_0) + v(t_0) - \int_{t_0}^{t} v^2\, ds \qquad t \geqq t_0 \tag{2.2.22}$$

$$\leqq \sigma(t) - \sigma(t_0) + v(t_0) \qquad t \geqq t_0 . \tag{2.2.23}$$

We can then proceed to the limit as $t \to \infty$ in (2.2.23) to obtain that

$$v(t) \to -\infty \qquad t \to \infty . \tag{2.2.24}$$

Arguing then as in Theorem 2.1.1, (2.2.24) leads to the contra diction (2.1.17). Thus no non-oscillatory solution can exist and this proves the theorem.

During the writing of these notes there appeared a

paper of Reid [50, p. 801] who also, independently, proved Theorem 2.2.3.

2.2A APPLICATIONS TO DIFFERENTIAL EQUATIONS:

THEOREM 2.2.1A:

Let $a(t)$ satisfy the conditions of Theorem 2.1.6A and suppose that $A(t) \geqq 0$ (where $A(t)$ is defined in (2.1.68)).

If

$$tA(t) \leqq \frac{1}{4} \qquad t \geqq t_0 \tag{2.2.25}$$

then (2.1.70) is non-oscillatory.

If $\varepsilon > 0$ is fixed and

$$tA(t) \geqq \frac{1}{4} + \varepsilon \qquad t \geqq t_0 \tag{2.2.26}$$

then (2.1.70) is oscillatory.

Proof: This is a consequence of Theorem 2.2.1 where $\sigma(t) \equiv A(t)$.

The first part of this theorem is due to Wintner [63, p. 260] and the second part follows almost immediately from this result (see [44, p. 131], [63, p. 259]).

<u>THEOREM 2.2.2A</u>:

Let $a(t)$ satisfy the conditions of Theorem 2.1.6A and suppose that $A(t) \geqq 0$.

If

$$\int_t^\infty A^2(s)ds \leqq \frac{1}{4} A(t) \qquad t \geqq t_0 \tag{2.2.27}$$

then (2.1.70) is non-oscillatory.

If $\varepsilon > 0$,

$$\int_t^\infty A^2(s)ds \geqq \left(\frac{1}{4} + \varepsilon\right) A(t) \qquad t \geqq t_0 \tag{2.2.28}$$

then (2.1.70) is oscillatory.

<u>Proof</u>: This follows from Theorem 2.2.2.

The above theorem is due to Opial [47, p. 309].

<u>THEOREM 2.2.3A</u>:

Let $a(t)$ be continuous on $[a, \infty)$ and suppose that (2.1.68) exists.

If

$$\int_a^\infty a(s)ds = \infty \tag{2.2.29}$$

then (2.1.70) is oscillatory.

Proof: We let $\sigma(t) = -\int_a^t a(s)ds$ in Theorem 2.2.3.

The latter theorem was proven by Fite [19, p. 347] in the case when $a(t) \geqq 0$ and was extended by Wintner [61, p. 115] for general $a(t)$.

§2.2B APPLICATIONS TO DIFFERENCE EQUATIONS:

THEOREM 2.2.1B:

Let the c_n, b_n satisfy the hypotheses of Theorem 2.1.4B and assume further that G_m, defined in the proof of Theorem 2.1.6B, is non-negative for $m \geqq m_0$.

If

$$\left\{ \sum_0^m \frac{1}{c_{n-1}} \right\} \sum_m^\infty (c_n + c_{n-1} + b_n) \leqq \frac{1}{4} \qquad m \geqq m_0 \qquad (2.2.30)$$

then (2.1.90) is non-oscillatory.

If

$$\left\{ \sum_0^m \frac{1}{c_{n-1}} \right\} \sum_m^\infty (c_n + c_{n-1} + b_n) \geqq \frac{1}{4} + \varepsilon \qquad m \geqq m_0 \qquad (2.2.31)$$

where $\varepsilon > 0$ is fixed, then (2.1.90) is oscillatory.

Proof: The first part is a consequence of Theorem 2.1.4B. An argument similar to the one used in the proof of the latter theorem shows that (2.2.3) is equivalent to (2.2.31).

As a consequence of this we obtain in particular,

COROLLARY 2.2.1B:

Let (b_n) be any sequence whose series is conditionally convergent and

$$\sum_m^\infty b_n \geqq 0 \qquad m \geqq m_0 \tag{2.2.32}$$

If

$$m \sum_m^\infty b_n \leqq \frac{1}{4} \qquad m \geqq m_0 \tag{2.2.33}$$

then (2.1.116) is non-oscillatory.

If $\varepsilon > 0$ is fixed and

$$m \sum_m^\infty b_n \geqq \frac{1}{4} + \varepsilon \qquad m \geqq m_0 \tag{2.2.34}$$

then (2.1.116) is oscillatory.

Proof: This follows from the discussion following Theorem 2.1.4B as applied to Theorem 2.2.1B.

Example 1: The discrete Euler equation,

$$\Delta^2 y_{n-1} + \frac{\gamma}{(n+1)^2} y_n = 0 \tag{2.2.35}$$

is oscillatory whenever $\gamma > \frac{1}{4}$. This is because

$$m \sum_{m}^{\infty} \frac{\gamma}{(n+1)^2} \geqq m\gamma \int_{m+1}^{\infty} x^{-2}\, dx$$

$$\geqq \frac{m}{m+1} \gamma$$

$$\geqq \frac{1}{4} + \varepsilon \qquad (2.2.36)$$

for some $\varepsilon > 0$ if m is sufficiently large since $\gamma > \frac{1}{4}$. Consequently the above corollary implies that (2.2.35) is oscillatory.

Using the discrete Euler equation (2.2.35) as a comparison equation we can deduce the following result. If (b_n) is a positive sequence such that, for fixed $\varepsilon > 0$,

$$(n+1)^2 b_n \geqq \frac{1}{4} + \varepsilon \qquad n \geqq n_0 , \qquad (2.2.37)$$

then (2.1.116) is oscillatory. For if we let

$$g_n = \left(\frac{1}{4} + \varepsilon\right)(n+1)^{-2} \qquad (2.2.38)$$

then $b_n \geqq g_n \geqq 0$ and Theorem 2.1.3B would lead immediately to a contradiction if (2.1.116) was assumed non-oscillatory.

Similarly it can be shown that if $b_n \geqq 0$ and

$$(n+1)^2 b_n \leqq \frac{1}{4} \qquad n \geqq n_0 \qquad (2.2.39)$$

then (2.1.116) is non-oscillatory.

THEOREM 2.2.2B:

Let (c_n) , (b_n) satisfy the hypotheses of Theorem 2.1.4B.

If for $m \geqq m_0$,

$$\sum_{i=m}^{\infty} \frac{1}{c_i} G_{i+1}^2 \leqq \frac{1}{4} G_{m+1} \tag{2.2.40}$$

then (2.1.103) is non-oscillatory.

If for $\varepsilon > 0$ fixed and $m \geqq m_0$,

$$\sum_{i=m}^{\infty} \frac{1}{c_i} G_{i+1}^2 \geqq \left(\frac{1}{4} + \varepsilon\right) G_m , \tag{2.2.41}$$

then (2.1.103) is oscillatory where

$$G_m = \sum_{i=m}^{\infty} (c_i + c_{i-1} + b_i) \tag{2.2.42}$$

Proof: The first part is Theorem 2.1.6B while the second part follows from the proof of the latter theorem. For (2.1.126) holds whenever $t \in [t_{m-1}, t_m)$. Thus letting $t \to t_m - 0$ in (2.1.126) we find

$$\begin{aligned} \int_{t_m}^{\infty} \sigma^2(s)ds &\geqq \sum_{m-1}^{\infty} \frac{1}{c_i} G_{i+1}^2 - \frac{1}{c_{m-1}} G_m^2 \\ &= \sum_{m}^{\infty} \frac{1}{c_i} G_{i+1}^2 \\ &\geqq \left(\frac{1}{4} + \varepsilon\right) G_m \end{aligned}$$

$$\geqq \left(\frac{1}{4} + \varepsilon\right)\sigma(t)$$

since $\sigma(t) = G_m$ when $t \in [t_{m-1}, t_m)$.

Thus

$$\int_t^\infty \sigma^2(s)ds \geqq \left(\frac{1}{4} + \varepsilon\right)\sigma(t) \qquad t \in [t_{m-1}, t_m)$$

and thus Theorem 2.2.2 applies. Consequently (2.1.103) is oscillatory.

The latter theorem gives the discrete analog of Opial's theorem (see Theorem 2.2.2A).

<u>THEOREM 2.2.3B</u>:

Let the (c_n) , (b_n) satisfy the hypotheses of Theorem 2.1.4B.

If

$$\sum_0^\infty (c_n + c_{n-1} + b_n) = \infty \tag{2.2.43}$$

then (2.1.103) is oscillatory.

<u>Proof</u>: We define $\sigma(t)$ as in (2.1.93) remembering that $\sigma(t)$ is a step-function with jumps at the (t_n) . From (2.1.94) we find that (2.2.43) is then equivalent to (2.2.21). Hence Theorem 2.2.3 applies and consequently (2.1.103) is oscillatory.

The latter theorem was shown by Hinton & Lewis [32, p. 426] and extended, in the same direction as Theorem 2.2.3, by Reid [50].

§2.3 AN OSCILLATION THEOREM IN THE NONLINEAR CASE:

In this section we extend a result of Butler [8, p. 75] which gives a necessary and sufficient condition for all continuable solutions of a second order nonlinear differential equation to be oscillatory. Our proof shall be mainly an adaptation of the proof in [8] to equations of the form

$$y'(t) = c - \int_a^t f\big(y(s)\big)\,d\sigma(s) \qquad (2.3.1)$$

where $t \in [a, \infty)$. A solution of (2.3.1) is again sought in the class of all locally absolutely continuous functions on $[a, \infty)$ whose first derivatives are locally of bounded variation on $[a, \infty)$.

We shall be mainly concerned with the "superlinear" case of (2.3.1): It is characterized by the convergence of the integral

$$\int^{\pm\infty} \frac{dt}{f(t)} \,. \qquad (2.3.2)$$

Equations of the form (2.3.1) include those which are of "Emden-Fowler type", i.e. those equations with

$$f(y) = y^{2n+1} \qquad n = 1, 2, \ldots . \qquad (2.3.3)$$

In the case of ordinary differential equations with f given as in (2.3.3) the result of Atkinson [2, p. 643] gave the first characterization of oscillatory solutions of the equation

$$y'' + p(t)y^{2n+1} = 0 \qquad n = 1, 2, \ldots , \qquad (2.3.4)$$

when $p(t) > 0$ and continuous, in terms of an integral condition on the coefficient. This has recently been generalized [8] to equations

$$y'' + p(t)f(y) = 0 \qquad (2.3.5)$$

where $p(t)$ is unrestricted to sign and f turns (2.3.5) into a "superlinear" equation. The result which we shall prove later on will give, in particular, a necessary and sufficient condition for the difference equation

$$\Delta^2 y_{n-1} + b_n f(y_n) = 0 \qquad n = 0, 1, \ldots \qquad (2.3.6)$$

to be oscillatory. As a corollary we shall obtain the discrete analog of Atkinson's theorem [2], i.e. If (b_n) is positive then, for $k \geq 1$,

$$\Delta^2 y_{n-1} + b_n y_n^{2k+1} = 0 \qquad n = 0, 1, \ldots$$

has a non-oscillatory solution if and only if

$$\sum_{0}^{\infty} n b_n < \infty . \tag{2.3.7}$$

In the following we shall assume that $f \in C'(-\infty, \infty)$ so that $f \circ y$ is absolutely continuous when y is a solution of (2.3.1), though it is possible to require that f be only absolutely continuous [8, p. 76]. In any case the integral in (2.3.1) has meaning as a Stieltjes integral.

THEOREM 2.3.1:

Let σ be a right-continuous function, locally of bounded variation and such that

$$P(t) \equiv \lim_{T\to\infty} \int_t^T d\sigma(s) \qquad \text{exists}, \tag{2.3.8}$$

and may be infinite. Suppose further that

a) $f \in C'(-\infty, \infty)$ and $yf(y) > 0$ for all $y \neq 0$. $f'(y) > 0$ if $y \neq 0$ and

$$\int_1^{\infty} \frac{dt}{f(t)} < \infty \qquad \int_{-\infty}^{-1} \frac{dt}{f(t)} < \infty ,$$

b) $\liminf_{T\to\infty} \int_t^T P(s)ds > -\infty$ for all t,

c) $\int_t^{\infty} \int_s^{\infty} P_-^2(r)drds < \infty$ where $P_{\mp}(t) = \max\{\mp P(t), 0\}$.

Then a necessary and sufficient condition for all solutions of (2.3.1), continuable over the half-axis, to be oscillatory is that

$$\int_t^\infty \left\{P(s) + \int_s^\infty P^2(r)\,dr\right\}ds = +\infty . \tag{2.3.9}$$

Note: We shall prove the sufficiency of (2.3.9) by first showing that (2.3.9) with $P^2(r)$ replaced by $P_+^2(r)$ will imply that all solutions oscillate (since (2.3.9) and (c) along with the relation $P^2(t) = P_+^2(t) + P_-^2(t)$ imply

$$\int_t^\infty \left\{P(s) + \int_s^\infty P_+^2(r)\,dr\right\}ds = +\infty \quad \Big). \tag{2.3.10}$$

<u>Proof:</u> (Sufficiency) Assume, on the contrary, that there is some non-oscillatory solution $y(t)$ which we can take to be positive (since $-y(t)$ is also a solution).

Thus

$$y(t) > 0 \qquad t \geqq t_0 . \tag{2.3.11}$$

We let

$$g(t) = \frac{y'(t)}{f[y(t)]} \qquad t \geqq t_0 \tag{2.3.12}$$

where the prime represents in general a right-derivative. Then $g(t)$ shall be right-continuous and locally of bounded variation on (t_0 , ∞) .

An integration by parts shows that, for $t_0 \leqq t \leqq T$,

$$\int_t^T \frac{1}{f(y(s))}\, dy'(s) = g(T) - g(t) + \int_t^T \frac{f'(y)(y')^2}{f(y)^2}\, ds \tag{2.3.13}$$

where we have omitted the variables in the integrand for simplicity. Moreover, an application of the integral equation (2.3.1) shows that

$$\int_t^T \frac{1}{f(y)}\, dy' = -\int_t^T d\sigma\ . \tag{2.3.14}$$

Hence combining (2.3.13-14) we find

$$g(t) = g(T) + \sigma(T) - \sigma(t) + \int_t^T f'(y)\left\{\frac{y'}{f(y)}\right\}^2 ds \tag{2.3.15}$$

whenever $t_0 \leqq t \leqq T$. Our basic assumption leads us to two cases:

$$\text{I)}\quad \limsup_{T\to\infty} \int_t^T P(s)ds = +\infty \qquad \text{some}\quad t \geqq t_0 \tag{2.3.16}$$

$$\text{II)}\quad \limsup_{T\to\infty} \int_t^T P(s)ds < \infty \qquad \text{all}\quad t \geqq t_0\ . \tag{2.3.17}$$

<u>Case I:</u> (2.3.16) implies that the relation is valid for all $t \geqq t_0$. If there is a sequence $T_n \uparrow \infty$ such that

$$g(T_n) \geqq 0 \tag{2.3.18}$$

then for n sufficiently large we shall have, for

$t_0 \leq t \leq T_n$

$$g(t) = g(T_n) + \sigma(T_n) - \sigma(t) + \int_t^{T_n} f'(y)g^2\,ds\,. \qquad (2.3.19)$$

Hence

$$g(t) \geqq \int_t^{T_n} d\sigma + \int_t^{T_n} f'(y)g^2\,ds \qquad (2.3.20)$$

so that letting $n \to \infty$ we obtain $g(t) \geq P(t)$. Taking the lim sup of both sides of the latter we obtain a contradiction on account of (a).

If no such sequence exists then we must have

$$g(t) < 0 \qquad t \geqq t_1 \qquad (2.3.21)$$

and so

$$y'(t) < 0 \qquad t \geqq t_1\,. \qquad (2.3.22)$$

(2.3.15) now implies

$$g(T) \leqq g(t) - \int_t^T d\sigma\,. \qquad (2.3.23)$$

Moreover (2.3.16) implies that $P(t) > 0$ for arbitrarily large t (not necessarily all such) which shows the existence of some $t_2 \geqq t_1$ such that

$$\int_{t_2}^T d\sigma > 0 \qquad T \geqq t_2 \qquad (2.3.24)$$

if we assume that $P(t) < \infty$ for large t . We note that

$P(t) = \infty$ for some t if and only if $\sigma(\infty) = \infty$ and thus $P(t) = \infty$ for all larger t. This would then imply that $g(t) \to -\infty$ as $t \to \infty$ because of (2.3.15). And we can then proceed as below.

Hence by (2.3.23) for $T \geqq t_2$,

$$g(T) \leqq g(t_2) \equiv -K < 0 . \tag{2.3.25}$$

Replacing t by t_2 in (2.3.15) and using (2.3.24 - 25) we obtain

$$g(T) \leqq -K + \int_{t_2}^{T} \frac{f'(y)|y'|}{f(y)} g \, ds . \tag{2.3.26}$$

If we write $\phi(t) = \dfrac{f'(y(t))|y'(t)|}{f(y(t))}$ then $\phi(t) > 0$ for $t \geqq t_2$ and so

$$g(T) \leqq -K + \int_{t_2}^{T} \phi(s) g(s) ds . \tag{2.3.27}$$

An application of the Gronwall inequality to (2.3.27) then gives

$$g(T) \leqq -K \frac{f(y(t_2))}{f(y(T))} \qquad T \geqq t_2 \tag{2.3.28}$$

i.e.

$$y'(T) \leqq -K f(y(t_2)) \qquad T \geqq t_2 \tag{2.3.29}$$

and the latter implies that $y(t)$ cannot remain positive for large t which contradicts (2.3.11).

Case II: In this case it is necessary that

$$\int_t^\infty \int_s^\infty P_+(r)\,dr\,ds = +\infty \qquad (2.3.30)$$

and that there be an $M_t > 0$ such that

$$\left| \int_t^T P(s)\,ds \right| \leqq M_t \qquad T \geqq t\ . \qquad (2.3.31)$$

We now proceed as in Case I. If there is a sequence $T_n \uparrow \infty$ such that (2.3.18) holds for large n we find from (2.3.20) that

$$g^2(t) \geqq P_+^2(t) \qquad t \geqq t_3\ . \qquad (2.3.32)$$

Now either i) $y(t) \geqq \delta > 0 \qquad t \geqq t_3$, or

ii) there is (t_n) such that $t_n \uparrow \infty$,

$y(t_n) \downarrow 0$.

i) Let $y(t) \geqq \delta > 0$, $t \geqq t_3$. Since

$$c = \inf\{f'(u) : \delta \leqq u < \infty\} > 0 \qquad (2.3.33)$$

(2.3.20) then implies that

$$g(t) \geqq P(t) + c \int_t^\infty g^2\,ds \qquad (2.3.34)$$

$$g(t) \geqq P(t) + c \int_t^{\infty} P_+^2(r)\,dr \,. \qquad (2.3.35)$$

Integrating both sides over $[t\,,T)$ and taking lim sups as $T \to \infty$ we get a contradiction to (a) since by hypothesis the integral of the right side of (2.3.35) is divergent.

ii) Let $t_n \uparrow \infty$ be such that $y(t_n) \downarrow 0$.

For large n and t fixed, $t \geqq t_3$,

$$0 \geqq \int_{y(t)}^{y(t_n)} \frac{ds}{f(s)} \geqq \int_t^{t_n} \left[P(s) + \int_s^{\infty} f'(y) P_+^2 dr \right] ds$$

because of (2.3.20), in the limit, and (2.3.32).

Thus

$$\int_{y(t)}^{y(t_n)} \frac{ds}{f(s)} \geqq \int_t^{t_n} P(s)\,ds \qquad (2.3.36)$$

so that

$$\liminf_{n\to\infty} \left\{ \int_{y(t)}^{y(t_n)} \frac{ds}{f(s)} \right\} \geqq \liminf_{n\to\infty} \int_t^{t_n} P(s)\,ds$$

and so

$$0 \leqq \int_0^{y(t)} \frac{ds}{f(s)} < \infty \qquad (2.3.37)$$

because of (b).

But

$$0 \leqq \int_0^{y(t)} \frac{f'(s)}{f(s)}\,ds \leqq \sup_{[0,y(t)]} f'(s) \cdot \int_0^{y(t)} \frac{ds}{f(s)} \ . \tag{2.3.38}$$

Because of (2.3.37) the expression on the right of (2.3.38) must be finite and thus

$$0 \leqq \int_0^{y(t)} \frac{f'(s)}{f(s)}\,ds < \infty \tag{2.3.39}$$

which is impossible since $f(0) = 0$. This contradiction settles this case.

If no such sequence T_n exists with (2.3.18) then $g(t) < 0$ for large t . Consequently $P(t) > 0$ for arbitrarily large t (or else $P(t) \leqq 0$ for $t \geqq T$ would imply $P_+(t) \equiv 0$ for $t \geqq T$ and this would contradict (2.3.30). We can then repeat the argument beginning at (2.3.24) and so obtain a contradiction.

(Necessity). We shall suppose that (2.3.9) is false and so set out to find a non-oscillatory solution of (2.3.1) using Schauder's fixed point theorem. First of all we let

$$R(t) = P(t) + \int_t^{\infty} P^2(s)\,ds \ .$$

Then

$$\left| \int_t^{\infty} R(s)\,ds \right| < \infty$$

and so

$$\int_t^\infty R_+ \, ds \quad , \quad \int_t^\infty R_- \, ds$$

are finite.

Moreover if $g \geqq 0$ then $(f+g)_+ \geqq f_+$ for any f . Using the latter we see that

$$\int_t^\infty R_+ \, ds \geqq \int_t^\infty P_+$$

$$\int_t^\infty R_- \, ds \geqq \int_t^\infty P_-$$

and so P is absolutely integrable. Thus

$$\int_t^\infty \left\{ |P(s)| + \int_s^\infty P^2 \right\} ds < \infty .$$

We define $Q(t)$ by

$$Q(t) = a|P(t)| + 2ab \int_t^\infty P^2 \, ds \tag{2.3.40}$$

where

$$a = \sup_{0 \leqq y \leqq 2} f(y)$$

$$b = \sup_{0 \leq y \leq 2} f'(y) .$$

We let B be the space of all bounded absolutely continuous functions on $[T, \infty)$, where T is to be chosen later, which have a right-derivative at each point that is bounded by some multiple of $Q(t)$.

$$B = \{y \in BAC[T,\infty) : y'_+ \equiv y' \text{ exists and } |y'(t)| \leqq (\text{const}) Q(t) \text{ for } t \geqq T\} .$$

Associate with B the norm defined, for $y \in B$, by

$$\|y\|_B = \|y\|_\infty + \left\|\frac{y'}{Q}\right\|_\infty$$

where $\| \ \|_\infty$ is the usual uniform norm. Then B is a Banach space.

For $n = 1, 2, \ldots$ we define a subset B_n of B by

$$B_n = \Big\{ y \in B : 0 \leqq y(t) \leqq 2 , \ t \geqq T , \ \|y'Q^{-1}\| \leqq 1 ,$$
$$|y'(t_2) - y'(t_1)| \leqq a|\sigma(t_2) - \sigma(t_1)| + |t_2 - t_1| , \ t_1, t_2 \in [T, T+n)$$
$$\text{and } y(t) = \text{constant on } [T+n, \infty) \Big\} .$$

For each n , B_n is a closed convex subset of B and in fact B_n is compact. The proof of the latter result is presented in Appendix II, Lemma II.1.1.

We continue as in [8] and define an operator A_n on B_n by

$$(A_n y)(t) = \begin{cases} 1 - \int_t^\infty \int_s^\infty f(y)\,d\sigma\,ds & t \in [T, T+n] \, . \quad (2.3.41) \\ \\ 1 - \int_{T+n}^\infty \int_s^\infty f(y)\,d\sigma\,ds & t \geqq T+n \, . \quad (2.3.42) \end{cases}$$

For $y \in B_n$, $t \in [T, T+n)$, $A_n y$ has a right-derivative at each t given by

$$(A_n y)'(t) = \int_t^\infty f\big(y(s)\big)\,d\sigma(s) \, .$$

If $y \in B_n$, then integration by parts shows that

$$\int_t^\tau f\big(y(s)\big)\,d\sigma(s) = f\big(y(t)\big)\int_t^\tau d\sigma + \int_t^\tau \left\{\int_s^\tau d\sigma\right\} f'(y)y'ds \, .$$

Hence

$$\int_t^\infty f(y)\,d\sigma = f\big(y(t)\big)P(t) + \int_t^\infty P(s)f'\big(y(s)\big)y'(s)\,ds \, .$$

Thus for $T \le t \le T+n$,

$$|(A_n y)'(t)| \leqq a|P(t)| + b\int_t^\infty |P(s)|\left\{a|P(s)| + 2ab\int_s^\infty P^2\right\}ds$$

and proceed to show as in [8] that

$$|(A_n y)'(t)| \leqq Q(t) \qquad t \in [T, T+n) \qquad (2.3.43)$$

if T is so large that

$$2b \int_t^{\infty} |P(s)|ds \; < 1 \qquad t \geqq T \; .$$

If $t \geqq T+n$, $(A_n y)'(t) = 0$ hence $A_n(B_n) \subset B$.

If, in addition, we require T so large that

$$\int_t^{\infty} Q(s)ds < 1 \qquad t \geqq T \tag{2.3.44}$$

then

$$0 \leqq (A_n y)(t) \leqq 2 \qquad t \geqq T \; , \tag{2.3.45}$$

since we can estimate the inner integrals in (2.3.41 - 42) by (2.3.43) and (2.3.44) then gives (2.3.45). For $y \in B_n$ it also follows from (2.3.42) that

$$(A_n y)(t) = \text{constant} \qquad t \geqq T+n \; .$$

If $T \leqq t_1 < t_2 < T+n$,

$$|(A_n y)'(t_2) - (A_n y)'(t_1)| = \left| \int_{t_1}^{t_2} f\big(y(s)\big) d\sigma(s) \right|$$

$$\leqq \left| f\big(y(t_1)\big) \int_{t_1}^{t_2} d\sigma \right| + \left| \int_{t_1}^{t_2} f'(y)y' \int_s^{t_2} d\sigma \, ds \right|$$

$$\leqq a \left| \int_{t_1}^{t_2} d\sigma \right| + \int_{t_1}^{t_2} b|Q(s)| \left| \int_s^{t_2} d\sigma \right| ds \; .$$

If necessary we can restrict T further by requiring that

$$b\,Q(s)\left|\int_s^{t_2} d\sigma\right| \leqq 1 \qquad t_2 \geqq s \geqq T\,.$$

Substituting this in the former equation we obtain

$$|(A_n y)'(t_2) - (A_n y)'(t_1)| \leqq a|\sigma(t_2) - \sigma(t_1)| + |t_2 - t_1|\,.$$

Thus $A_n(B_n) \subset B_n$.

There remains to show that A_n is continuous: This can be done as in [8, p. 82] with the appropriate modifications in the definitions of $a(\delta)$, $b(\delta)$ there.

i.e.

$$a(\delta) = \sup\{|f(y) - f(x)| : 0 \leqq x, y \leqq 2,\ |y - x| < \delta\}$$

$$b(\delta) = \sup\{|f'(y) - f'(x)| : 0 \leqq x, y \leqq 2,\ |y - x| < \delta\}\,.$$

From the above definitions we see that as $\delta \to 0$ both $a(\delta)$, $b(\delta) \to 0$ since $f \in C'$.

If δ is chosen sufficiently small so that, for given ε , $\|x - y\| < \delta$,

$$2\,c(\delta) < \varepsilon$$

where $c(\delta) = \max\{a(\delta), 2ab(\delta) + \delta b(a+1)\}$, then

$$\|A_n y - A_n x\| < \varepsilon$$

which shows that A_n is continuous. The Schauder fixed point theorem therefore implies that there is some $x_n \in B_n$ such

$$A_n x_n = x_n$$

and this x_n is a solution of (2.3.41 - 42).

Since

$$|x_n(t_2) - x_n(t_1)| \leqq \int_{t_1}^{t_2} |x_n'(t)|dt \leqq \int_{t_1}^{t_2} Q(t)dt$$

the family $\{x_n\}$ is equicontinuous and uniformly bounded hence there is a subsequence which converges uniformly on compact intervals to a non-negative function which satisfies the integral equation and is eventually non-oscillatory.

Note: Since

$$\int_t^{\infty} Q(s)ds < \infty$$

then $\lim \sigma(t)$ (exists and) must be finite and so we can assume it is zero. Thus $Q(t) \to 0$ as $t \to \infty$.

If we assume that $\sigma \in C'(a, \infty)$,

$$\sigma(t) = \int_a^t p(s)ds \qquad t \geqq a \tag{2.3.46}$$

then (2.3.1) becomes, upon differentiation,

$$y''(t) + p(t)f(y(t)) = 0 \tag{2.3.47}$$

and

$$P(t) = \lim_{T\to\infty} \int_t^T p(s)ds .$$

Thus Theorem 3.1.1 includes the theorem of Butler. For the various applications of the latter theorem to differential equations we shall refer to [8]. We shall be mainly concerned with formulating oscillation criteria for nonlinear difference equations.

Defining σ as in (1.1.25) with b_n replaced by $-b_n$ and the (c_n) is positive and satisfies (2.1.91) we can specify the sequence (t_n) by (2.1.92) with $t_{-1} = a$. It can then be shown using the methods of Chapter 1 that the equation (2.3.1) will have a solution $y(t)$ whose values $y(t_n) \equiv y_n$ satisfy the nonlinear recurrence relation

$$c_n y_{n+1} + c_{n-1}y_{n-1} - (c_n + c_{n-1})y_n + b_n f(y_n) = 0. \quad (2.3.48)$$

With $\sigma(t)$ defined above we see that, for $m \geqq 1$,

$$\sigma(t) = \sigma(a) + \sum_0^{m-1} b_n \qquad t \in [t_{m-1}, t_m) \quad (2.3.49)$$

so that if $t \in [t_{m-1}, t_m)$,

$$P(t) = \sum_m^\infty b_n \equiv P_{m-1} . \quad (2.3.50)$$

THEOREM 2.3.2:

Let (c_n) , (b_n) be as above.

Assume the following:

a) That f should satisfy the assumption (a) of Theorem 3.1.1.

b) $\liminf_{N\to\infty} \sum_{n=m}^{N} \frac{1}{c_n} P_n > -\infty$.

c) Let $Q_{j-1} \equiv \max\{-P_{j-1}, 0\}$ and

$$R_{j-1} \equiv \sum_{i=j-1}^{\infty} \frac{1}{c_i} Q_i^2 .$$

Then $\sum_{j=n}^{\infty} \frac{1}{c_j} R_j < \infty$.

Under the above hypotheses we have that a necessary and sufficient condition for (2.3.48) to be oscillatory is that

$$\sum_{m=n}^{\infty} \frac{1}{c_{m-1}} \left\{ P_{m-1} + \sum_{i=m-1}^{\infty} \frac{1}{c_i} P_i^2 \right\} = +\infty . \qquad (2.3.51)$$

Proof: This follows from the preceding theorem with a suitable interpretation of the hypotheses and the integral condition (2.3.9) which becomes (2.3.51).

COROLLARY 2.3.1:

Let $P_n > 0$ for $n \geqq N$. Then a necessary and

sufficient condition for (2.3.48) to be oscillatory is that

$$\sum_{m=n}^{\infty} \frac{1}{c_{m-1}} P_{m-1} + \sum_{m=n}^{\infty} \frac{1}{c_{m-1}} \sum_{i=m-1}^{\infty} \frac{1}{c_i} P_i = +\infty . \qquad (2.3.52)$$

Proof: This follows immediately from Theorem 2.3.2 because (2.3.51 - 52) are equivalent since $P_n > 0$.

THEOREM 2.3.3:

Let σ satisfy the basic hypotheses of Theorem 2.3.1.

a) If

$$\lim_{t\to\infty} \sigma(t) = \infty \qquad (2.3.53)$$

then (2.3.1) is oscillatory.

b) If $\sigma(t)$ is non-decreasing then a necessary and sufficient condition for (2.3.1) to be oscillatory is that

$$\int_{t_0}^{\infty} t d\sigma(t) = \infty . \qquad (2.3.54)$$

Proof: a) follows immediately from Theorem 2.3.1 since $P(t) \equiv \infty$ for all t . Hence (2.3.9) is identically satisfied. To prove b) we must show that (2.3.9) is equivalent to (2.3.54).

If σ is non-decreasing and (2.3.9) is finite then

$$\int_{t}^{\infty} P(s)ds < \infty . \qquad (2.3.55)$$

Consequently $\sigma(t)$ must tend to a finite limit, which we can assume is zero. Thus $P(t) = -\sigma(t)$ and the latter is non-increasing.

On the other hand if (2.3.55) holds then

$$\int_{t_0}^{\infty} \left\{P(t) + \int_t^{\infty} P^2\right\}dt < \int_{t_0}^{\infty} \left\{P(t) + P(t) \int_t^{\infty} P(s)ds\right\} dt$$

since P is non-increasing,

$$< \int_{t_0}^{\infty} P(t)\left\{1 + \int_t^{\infty} P(s)ds\right\} dt .$$

For t sufficiently large

$$1 + \int_t^{\infty} P(s)ds = O(1) . \qquad (2.3.56)$$

Hence (2.3.9) is finite. Thus we have shown that (2.3.9) is equivalent to (2.3.55).

Now

$$\int_{t_0}^{\infty} P(t)dt = \int_{t_0}^{\infty} \int_t^{\infty} d\sigma(s)dt$$

$$= \int_{t_0}^{\infty} \left(\int_{t_0}^{t} ds\right) d\sigma(t)$$

$$= \int_{t_0}^{\infty} (t - t_0)d\sigma(t) \qquad (2.3.57)$$

with the interchange justified by the Fubini theorem. Thus

(2.3.55) is finite if and only if (2.3.57) is finite and the latter is finite if and only if (2.3.54) is finite. Thus (2.3.9) and (2.3.54) must diverge together and this completes the proof.

COROLLARY 2.3.2:

Let (c_n) be positive and satisfy (2.1.91). Let the sequence (t_n) satisfy (2.1.92) with $t_{-1} = a$. Suppose that the sequence (b_n) is also positive.

a) The necessary and sufficient condition for all the solutions of (2.3.48) to be oscillatory is that

$$\sum_{m=n_0}^{\infty} \left\{a + \sum_{0}^{m} \frac{1}{c_{i-1}}\right\} b_m = \infty . \qquad (2.3.58)$$

b) If

$$\sum_{m=n_0}^{\infty} b_m = \infty \qquad (2.3.59)$$

then all solutions of (2.3.48) are oscillatory (here (b_n) is not necessarily positive).

Proof: Part (b) follows from (a) of Theorem 2.3.3 where $\sigma(t)$ satisfies (2.3.49).

Part (a) follows from (b) of Theorem 2.3.3 where we only need to note that $\sigma(t)$ is as in (2.3.49) and

$$t_m = t_{-1} + \sum_{0}^{m} \frac{1}{c_{i-1}} \qquad m \geq 0 ,$$

so that (2.3.58) is equivalent to (2.3.54).

In particular we can choose $a = -1$ and $c_n = 1$ for all $n = -1, 0, 1, \ldots$. We then obtain from (2.3.48) that, when $b_n > 0$,

$$\Delta^2 y_{n-1} + b_n f(y_n) = 0 \tag{2.3.60}$$

is oscillatory if and only if

$$\sum_{m=n_0}^{\infty} m b_m = \infty . \tag{2.3.61}$$

This is the discrete analog of Atkinson's theorem [2] which follows from the previous corollary.

Example 1: Let $c_n = n+2$, $n = -1, 0, 1, \ldots$ and let $b_n = 1/(n+1)$, $n = 0, 1, \ldots$. Then (2.1.91) is satisfied and if we choose $a = 0$, then

$$\sum_0^{\infty} \left\{ \sum_0^{m} \frac{1}{c_{i-1}} \right\} b_m = \sum_0^{\infty} \left\{ 1 + \frac{1}{2} + \cdots + \frac{1}{m+1} \right\} \frac{1}{m+1}$$

$$> \sum_0^{\infty} \frac{1}{m+1} = \infty .$$

Hence Corollary 2.3.2(a) implies that all solutions of (2.3.48) are oscillatory where f is any function satisfying (a) of Theorem 2.3.1.

Example 2: If we let (c_n) be as in Example 1 above and

$$b_n = \frac{1}{(n+1)^{1+\delta}} \qquad \delta > 0$$

then

$$\sum_{0}^{\infty} \left\{ \sum_{0}^{m} \frac{1}{c_{i-1}} \right\} b_m < \infty$$

so that (2.3.48) has at least one nontrivial non-oscillatory solution.

For

$$\sum_{m=0}^{\infty} b_m \sum_{i=0}^{m} \frac{1}{c_{i-1}} = \sum_{i=0}^{\infty} \frac{1}{c_{i-1}} \sum_{m=i}^{\infty} b_m$$

since the series have positive terms and

$$\sum_{i=0}^{\infty} \frac{1}{i+1} \sum_{m=i}^{\infty} \frac{1}{(m+1)^{1+\delta}} < \sum_{i=1}^{\infty} \frac{1}{i+1} \int_{i-1}^{\infty} (x+1)^{-1-\delta}\, dx + \sum_{m=0}^{\infty} (m+1)^{-1-\delta} \qquad (2.3.62)$$

i.e.

$$< \delta^{-1} \sum_{i=1}^{\infty} \frac{1}{i^{\delta}(i+1)} + O(1)$$

$$< \delta^{-1} \sum_{i=1}^{\infty} \frac{1}{i^{\delta+1}} + O(1)$$

and since $\delta > 0$ the last term is finite.

<u>Example 3</u>: Let $c_n = 1$ and let $b_n = (-1)^n/n$, $n = 1, 2, \ldots$.

Then

$$P_{k-1} = \sum_{n=k}^{\infty} \frac{(-1)^n}{n} = (-1)^k \int_0^1 \frac{t^{k-1}}{1+t}\,dt \qquad (2.3.63)$$

and thus P_{k-1} is positive if k is even and negative if k is odd. By the alternating series theorem it is readily verified that

$$\lim_{N\to\infty} \sum_{k=n}^{N} P_{k-1} \qquad \text{exists}$$

and is finite. (Its value can be computed via 2.3.63.) Thus (b) of Theorem 2.3.2 is verified.

Now

$$\sum_{m=n}^{\infty} \left\{ \sum_{k=m-1}^{\infty} S_k^2 \right\} = \sum_{k=n-1}^{\infty} \sum_{m=n}^{k+1} S_k^2$$

where

$$S_k = \max\{P_k , 0\} .$$

Hence

$$\sum_{m=n}^{\infty} \left\{ \sum_{k=m-1}^{\infty} S_k^2 \right\} = \sum_{k=n-1}^{\infty} (k-n+2) S_k^2 .$$

A look at (2.3.63) shows that $\sum S_k^2 < \infty$. Moreover,

$$\sum_{k=n-1}^{\infty} kS_k^2 = \sum_{k=n-1}^{\infty} k\left\{\int_0^1 \frac{t^k}{1+t}\,dt\right\}^2$$

$$> \sum_{k=n-1}^{\infty} k\cdot\frac{1}{4}\left\{\int_0^1 t^k\,dt\right\}^2 = \infty .$$

Hence

$$\sum_{m=n}^{\infty}\left\{\sum_{k=m-1}^{\infty} S_k^2\right\} = \infty$$

and so Theorem 2.3.2 implies that

$$\Delta^2 y_{n-1} + \frac{(-1)^n}{n} f(y_n) = 0$$

is oscillatory.

Example 4: Let $c_n = 1$, $b_n = (-1)^n/n^{1+\delta}$, $\delta > 0$, $n = 1, 2, \ldots$.

Then

$$P_{n-1} = \sum_{m=n}^{\infty} \frac{(-1)^m}{m^{1+\delta}} < \infty$$

and the alternating series theorem implies that

$$\left|\sum_{m=n}^{\infty} P_{m-1}\right| < \infty .$$

As in the previous example

$$\sum_{m=n}^{\infty} \left\{ \sum_{k=m-1}^{\infty} P_k^2 \right\} = \sum_{k=n-1}^{\infty} (k - n + 2) P_k^2$$

and

$$\sum_{k=n-1}^{\infty} kP_k^2 = \sum_{k=n-1}^{\infty} k \left| \sum_{m=k-1}^{\infty} \frac{(-1)^m}{m^{1+\delta}} \right|^2$$

$$< \sum_{k=n-1}^{\infty} k \cdot \frac{1}{(k-1)^{2+2\delta}}$$

$$= \sum_{k=n-1}^{\infty} \left\{ 1 + \frac{1}{k-1} \right\} \frac{1}{(k-1)^{1+2\delta}}$$

$$< \infty .$$

Thus $\sum (k - n + 2) P_k^2 < \infty$ and so (2.3.51) is finite. Thus the corresponding equation has at least one nontrivial non-oscillatory solution.

ADDENDA: ON A RELATION BETWEEN NON-OSCILLATORY DIFFERENTIAL AND DIFFERENCE EQUATIONS

The preceding chapter shows the similarities which arise when one studies the oscillatory and non-oscillatory behaviour of differential and difference equations. It seems plausible that a general theorem exists which reduces the problem of determining when all solutions to a given differential equation

$$y'' + f(t)y = 0 \qquad (2.3.64)$$

where, say, f is continuous on $[a, \infty)$, are oscillatory, to the same problem but for a difference equation

$$\Delta^2 y_{n-1} + f_n y_n = 0 . \qquad (2.3.65)$$

One such theorem is the following

THEOREM 2.3.4:

Let f be a continuous non-increasing function on $[a, \infty)$.

Let $t_{-1} \geq a$, and

$$t_n - t_{n-1} = 1 \qquad n = 0, 1, \ldots . \qquad (2.3.66)$$

Denote $f(t_n)$ by f_n.

Then a necessary and sufficient condition that (2.3.64) be

non-oscillatory is that the difference equation (2.3.65) be non-oscillatory.

Proof: The idea is to rewrite both as Stieltjes integral equations and use Theorem 2.1.1. Since f is non-increasing there is no loss of generality in assuming that $f \geqq 0$ on $[a, \infty)$. Let σ_1, σ_2 be defined by

$$\sigma_1(t) = \sum_{m-1}^{\infty} f_n \qquad t \in [t_{m-1}, t_m) \tag{2.3.67}$$

$$\sigma_2(t) = \sum_{m+1}^{\infty} f_n \qquad t \in [t_{m-1}, t_m) . \tag{2.3.68}$$

If we let $y(t)$, $z(t)$ be solutions of

$$y'(t) = c_1 + \int_a^t y d\sigma_1 \tag{2.3.69}$$

$$z'(t) = c_2 + \int_a^t z d\sigma_2 \tag{2.3.70}$$

then $y(t_n) \equiv y_n$, $z(t_n) \equiv z_n$ will satisfy the difference equations

$$\Delta^2 y_{n-1} + f_{n-1} y_n = 0 \tag{2.3.71}$$

$$\Delta^2 z_{n-1} + f_{n+1} z_n = 0 . \tag{2.3.72}$$

Now a change of variable shows that (2.3.65) is non-oscillatory if and only if either of (2.3.71) or (2.3.72) is non-oscillatory.

Suppose that (2.3.64) is non-oscillatory. Then Theorem 2.1.1 implies that the integral equation

$$v(t) = \int_t^\infty f(s)ds + \int_t^\infty v^2(s)ds \qquad (2.3.73)$$

has a solution for sufficiently large t and so f must be integrable at ∞. Consequently (2.3.67 - 68) are both finite. Let $t \in [t_{m-1}, t_m)$. Estimating the integral by upper and lower sums we obtain

$$\sum_{m+1}^\infty f_n \leqq \int_t^\infty f(s)ds \leqq \sum_{m-1}^\infty f_n \qquad t \in [t_{m-1}, t_m) .$$

(2.3.74)

In particular

$$\int_t^\infty f(s)ds \geqq \sigma_2(t) \geqq 0 \qquad t \in [t_{m-1}, t_m) \qquad (2.3.75)$$

and so this is true for all t.

Since (2.3.73) has a solution Corollary 2.1.2 applies, because of (2.3.75), and so

$$v(t) = \sigma_2(t) + \int_t^\infty v^2\, ds$$

has a solution for sufficiently large t. Hence (2.3.70) is non-oscillatory and consequently so is (2.3.72). Hence (2.3.65) is non-oscillatory.

If now we assume that (2.3.65) is non-oscillatory then the same must be true of (2.3.71). Theorem 2.1.1 therefore implies that

$$v(t) = \sigma_1(t) + \int_t^\infty v^2\, ds$$

has a solution for large t. But (2.3.74) implies that

$$\sigma_1(t) \geqq \int_t^\infty f(s)\,ds \qquad t \in [t_{m-1}, t_m) \; . \tag{2.3.76}$$

Hence the latter holds for all t and thus applying Corollary 2.1.2 again we find that (2.3.73) has a solution and thus (2.3.64) has a non-oscillatory solution which implies that it is non-oscillatory. This completes the proof.

As a consequence we immediately obtain that the discrete Euler equation

$$\Delta^2 y_{n-1} + \frac{\gamma}{(n+1)^2}\, y_n = 0 \tag{2.3.77}$$

is oscillatory when $\gamma > \frac{1}{4}$ and non-oscillatory when $\gamma \leqq \frac{1}{4}$ (see example 1, section 2.1, and Example 1 of section 2B).

Furthermore we shall have [30, p. 30],

$$\Delta^2 y_{n-1} + \lambda y_n = 0 \qquad n = 0, 1, \ldots$$

non-oscillatory when $\lambda \leq 0$ and oscillatory when $\lambda > 0$

because of the analogous property for

$$y'' + \lambda y = 0 .$$

CHAPTER 3

INTRODUCTION:

The purpose of this chapter is to provide a basic framework for the theory of operators generated by the Volterra-Stieltjes integral equations encountered in the preceding chapters. The method used here will show that these integral equations can be thought of as defining generalized differential operators. Such a formalism was undertaken by I.S. Kac [35] though the application there was only to differential equations. A different formalism has been used by H. Langer [41] to deal with the notion of an operator defined by a Stieltjes integral equation of the form (2.1.0). The method which we shall use here is a natural extension of that used by Kac [35] and its applications will include differential equations and in particular, Sturm-Liouville problems with indefinite weight functions and difference equations.

In section 2 we shall proceed to define the generalized differential expression

$$\ell[f] = -\frac{d}{d\nu(x)}\left(f'(x) - \int_a^x f(s)\,d\sigma(s)\right) \qquad (3.0.0)$$

where ν, σ are real right-continuous functions of bounded variation, after having given the background material in section 1.

In section 3 we shall study the Weyl classification (limit-point, limit-circle) of singular generalized differential operators with an application to the particular case

$$-y'' + q(t)y = \lambda r(t)y \qquad t \in [a, \infty) \tag{3.0.1}$$

where the weight-function $r(t)$ vanishes identically on some interval. Other applications will include the three-term recurrence relation

$$-c_n y_{n+1} - c_{n-1} y_{n-1} + b_n y_n = \lambda a_n y_n \tag{3.0.2}$$

where $c_n > 0$. These will be discussed in section 4.

In section 5 we give some criteria which can be used to determine whether a certain equation is in the limit-point or limit-circle case. In section 6 we shall be considering the self-adjointness and, more generally, the J-self-adjointness of such generalized operators.

In section 7 we discuss the finiteness of Dirichlet integrals associated with (3.0.0) and consider the chain of implications [39]

$$\text{DI} \Rightarrow \text{CD} \Rightarrow \text{SLP} \Rightarrow \text{LP}$$

where these abbreviations stand for Dirichlet, Conditionally Dirichlet, Strong Limit-Point, Limit-Point respectively.

Finally, in section 8 we define these notions for a three-term recurrence relation and give some examples.

§3.1 GENERALIZED DERIVATIVES:

Let μ, ν be two real right-continuous functions locally of bounded variation on $[a, \infty)$, $a > -\infty$. Then at each interior point $\gamma \in \mathbb{R}$,

$$\lim_{x \to \gamma \pm 0} \mu(x) \quad , \quad \lim_{x \to \gamma \pm 0} \nu(x)$$

both exist and are finite.

Associated with μ (or ν) is a set function m_μ defined on intervals $(\alpha, \beta]$ and $[\alpha, \beta]$ in $[a, \infty)$ by

$$m_\mu(\alpha, \beta] = \mu(\beta) - \mu(\alpha) \qquad \alpha \leqq \beta \tag{3.1.0}$$

$$m_\mu[\alpha, \beta] = \mu(\beta) - \mu(\alpha - 0) \quad \alpha \leqq \beta . \tag{3.1.1}$$

When μ is non-decreasing then μ induces a σ-finite Borel measure on $[a, \infty)$ [55, p. 262]. Since every function μ of bounded variation is a difference of two non-decreasing functions such a function will induce a σ-finite signed Borel measure on $[a, \infty)$ [55, p. 264, ex. 11] which is finite if the original function is bounded on $[a, \infty)$. We will denote such a measure by m_μ . Then, for every Borel set E ,

$$m_\mu(E) = m_\mu^+(E) - m_\mu^-(E) \qquad (3.1.2)$$

where m_μ^+ , m_μ^- are the positive and negative variations of μ obtained by its Jordan decomposition [24, p. 123]. Thus each function μ right-continuous and locally of bounded variation induces a σ-finite signed Borel measure on $[a, \infty)$ satisfying (3.1.2). The measure $|m_\mu|$ defined by

$$|m_\mu|(E) = m_\mu^+(E) + m_\mu^-(E) \qquad (3.1.3)$$

where m_μ^+ , m_μ^- are as in (3.1.2) is called the total variation (or total variation measure) of μ .

If μ , ν are signed measures we say that μ is absolutely continuous with respect to ν if $|m_\mu|(E) = 0$ for every measurable set E for which $|m_\nu|(E) = 0$ [24, p. 125]. If $\mu(x)$ is ν-absolutely continuous there exists a finite-valued measurable function ϕ such that

$$m_\mu(E) = \int_E \phi \, dm_\nu \qquad (3.1.4)$$

for every Borel set E [24, p. 131, ex. 4]. The function ϕ appearing in (3.1.4) is called the Radon-Nikodym derivative of μ with respect to ν . It is unique in the sense that if ψ is another measurable function with this property then $\phi = \psi$

ν-almost everywhere (that is they are equal everywhere except possibly on a set E with $|m_\nu|(E) = 0$).

When $\mu(x)$ is non-decreasing and right-continuous and ϕ is a non-negative Borel measurable function the Lebesgue-Stieltjes integral of ϕ with respect to μ is defined by

$$\int \phi(x)d\mu(x) \equiv \int \phi dm_\mu . \tag{3.1.5}$$

If ϕ is both positive and negative it is integrable with respect to μ if it is integrable with respect to m_μ. When $\mu(x)$ is of bounded variation

$$\int_a^b \phi(x)d\mu(x)$$

agrees with the ordinary Riemann-Stieltjes integral whenever the latter is defined [55, p. 261].

When μ, ν are of bounded variation and μ is ν-absolutely continuous,

$$\frac{d\mu}{d\nu}(x) \equiv \lim_{h\downarrow 0} \left\{\frac{\mu(x+h) - \mu(x-h)}{\nu(x+h) - \nu(x-h)}\right\} \tag{3.1.6}$$

exists m_ν-almost everywhere, in particular when x is a point of discontinuity of ν it exists and

$$\frac{d\mu}{d\nu}(x) = \frac{\mu(x+0) - \mu(x-0)}{\nu(x+0) - \nu(x-0)}$$

$$= \phi(x)$$

where ϕ is the Radon-Nikodym derivative of μ with respect to ν defined in (3.1.4). (For general information on these derivatives, see [24, p. 132].)

From (3.1.4-5), for $\alpha, \beta \in [a, \infty)$, [18, p. 134],

$$\mu(\beta \pm 0) - \mu(\alpha \pm 0) = \int_{\alpha\pm 0}^{\beta\pm 0} \phi(x)\,d\nu(x)$$

$$= \int_{\alpha\pm 0}^{\beta\pm 0} \frac{d\mu(x)}{d\nu(x)} \cdot d\nu(x) .$$

When μ, ν are of bounded variation and have no common points of discontinuity then

$$\int_{\alpha\pm 0}^{\beta\pm 0} \mu(x)\,d\nu(x) = [\mu(x)\nu(x)]_{\alpha\pm 0}^{\beta\pm 0} - \int_{\alpha\pm 0}^{\beta\pm 0} \nu(x)\,d\mu(x) .$$

This is the general formula for integration by parts. When x is a single point its μ (or ν) measure shall be denoted by $\mu\{x\}$ (or $\nu\{x\}$). It is defined by (3.1.1).

§3.2 GENERALIZED DIFFERENTIAL EXPRESSIONS OF THE SECOND ORDER:

In this section we shall essentially pursue the approach of Kac [35] in the definition of a generalized differential expression of the second order on some interval I, i.e. for $x \in I$, $a \in I$ fixed,

$$\ell[y](x) \equiv -\frac{d}{d\nu(x)}\left\{y'_+(x) - \int_{a+0}^{x+0} y(s)\,d\sigma(s)\right\} \qquad (3.2.0)$$

where σ was assumed to be locally of bounded variation on I and ν was non-decreasing.

It shall be convenient to assume that σ is, in addition, right-continuous and ν an arbitrary function of bounded variation. We shall see below that it is still possible, in the latter case, to define (3.2.0).

Basic assumptions:

Throughout the remainder of this chapter we shall assume that:

a) ν, σ both have at most a finite number of discontinuities in finite intervals (see the hypotheses of §1.1).

b) If I is a finite interval then both ν, σ shall be continuous at the end-points of I . If I is a semi-infinite interval then both ν, σ shall be continuous at the finite end, and that neither has a discontinuity at infinity, i.e.

$$\lim_{x\to\infty} \nu(x) \quad , \quad \lim_{x\to\infty} \sigma(x)$$

both should exist (may be infinite).

Thus both a), b) shall be assumed in addition to the usual hypotheses of right-continuity and bounded variation for ν, σ .

Let μ , ν be two right-continuous functions of bounded variation. As we saw in the previous section each of these induces a σ-finite signed Borel measure on I . If, in addition, we assume that m_μ is absolutely continuous with respect to m_ν then the Radon-Nikodym derivative

$$\phi \equiv \frac{dm_\mu}{dm_\nu} \qquad (3.2.1)$$

exists ν-almost everywhere and we have relation (3.1.4). Moreover (3.2.1) agreees with (3.1.6) ν-almost everywhere.

Let ν , σ be two right-continuous functions of bounded variation on $[a , b]$.

A function f is said to belong to the class $\mathcal{D}_\nu \equiv \mathcal{D}_\nu[a , b]$ if

i) f is absolutely continuous on $[a , b]$,

ii) f has at each point $x \in [a , b]$ a right-derivative $f_+'(x)$ and the function

$$\mu(x) \equiv f_+'(x) - \int_a^x f(s)d\sigma(s)$$

is ν-absolutely continuous on $[a , b]$.

We note that (ii) necessitates that $f_+'(x)$ be of bounded variation on $[a , b]$. The quantity $f_+'(b)$ can be termed an "associated number" [35, p. 212].

Thus the preceding discussion shows that if $f \in \mathcal{D}_\nu$ the quantity

$$\ell[f](x) \equiv -\frac{d}{d\nu(x)}\left\{f'_+(x) - \int_a^x f(s)d\sigma(s)\right\} \qquad (3.2.2)$$

exists ν-almost everywhere on $[a , b]$ (i.e., it has meaning everywhere except possibly on a set on which the total variation of ν is zero).

A particular case of a generalized differential expression is (2.1.0). To see this we let $\nu(x) = x$ so that m_ν is Lebesgue measure and let $y(x)$ be any solution of (2.1.0). It is clear that

$$\frac{d}{dx}\left\{y'(x) - \int_a^x y(s)d\sigma(s)\right\} = 0$$

so that equations of the form (2.1.0) can be brought into the form (3.2.2).

By a *solution* of the generalized differential equation

$$-\frac{d}{d\nu(x)}\left\{f'(x) - \int_a^x f(s)d\sigma(s)\right\} = \phi(x) \qquad (3.2.3)$$

is meant a function $f \in \mathcal{D}_\nu$ which satisfies (3.2.3) ν-almost everywhere, which we shall abbreviate as $[\nu]$. The real purpose for defining expressions of the form (3.2.2) is contained in the following theorem.

<u>THEOREM 3.2.0</u>:

In order that a function $f(x)$ be a solution of (3.2.3) on $[a, b]$, say, it is necessary and sufficient that it satisfy the integral equation

$$f(x) = \alpha + \beta(x-a) + \int_a^x (x-s)f(s)\,d\sigma(s) - \int_a^x (x-s)\phi(s)\,d\nu(s) \tag{3.2.4}$$

for some α, β, $x \in [a, b]$.

<u>Proof</u>: This is not very different from [35, p. 215]. For let f be a solution of (3.2.4). Then f is absolutely continuous on $[a, b]$ and f has a right-derivative for each x, given by

$$f'(x) = \beta + \int_a^x f(s)\,d\sigma(s) - \int_a^x \phi(s)\,d\nu(s) . \tag{3.2.5}$$

i.e.

$$\mu(x) = \beta - \int_a^x \phi(s)\,d\nu(s) \qquad x \in [a, b]$$

where μ was defined in (ii). The latter shows that μ is ν-absolutely continuous and hence

$$\phi(x) = -\frac{d\mu}{d\nu}(x) = -\frac{d}{d\nu(x)}\left\{f'(x) - \int_a^x f\,d\sigma\right\} \qquad [\nu] . \tag{3.2.6}$$

Consequently $f \in \mathcal{D}_\nu$ and f satisfies (3.2.3) $[\nu]$.

On the other hand if f is a solution of (3.2.3), $x \in [a, b]$,

$$\int_a^x \phi(s)\,d\nu(s) = -\int_a^x \frac{d}{d\nu(s)}\left\{f'(s) - \int_a^s f\,d\sigma\right\}d\nu(s)$$

$$= -\int_a^x d\left\{f'(s) - \int_a^s f\,d\sigma\right\}$$

from a result in [24, p. 134],

$$= -f'(x) + \int_a^x f(s)\,d\sigma(s) + f'(a)$$

which is (3.2.5).

Integrating the latter over $[a, x]$ we obtain (3.2.4). The proof is now complete.

<u>THEOREM 3.2.1</u>: [35, p. 250]

If $f, g \in \mathcal{D}_\nu$ $[\alpha, \beta] \subset I$,

$$\int_\alpha^\beta \{f(s)\overline{\ell[g](s)} - \overline{g(s)}\ell[f](s)\}d\nu(s) = [f'\bar{g} - f\bar{g}']_\alpha^\beta .$$

This is the *Lagrange identity*.

<u>Proof</u>: This is exactly Lemma 5, p. 250 of [35] though it is shown there when ν is non-decreasing. The proof carries through without any essential change when ν is of bounded variation and so is omitted.

Thus solutions to the initial value problem

$$-\frac{d}{d\nu(x)}\left\{f'(x) - \int_a^x f d\sigma\right\} = \phi \tag{3.2.7}$$

$$f(a) = \alpha$$

$$f'(a) = \beta \tag{3.2.8-9}$$

exist on $[a, b]$ because solutions to (3.2.4) exist as we saw in Appendix I.

Note: If $I = [a, \infty)$ and $p(x) > 0$ is right-continuous, locally of bounded variation and

$$\int^{\infty} \frac{dx}{p(x)} = \infty$$

the theory which would arise by consideration of the more general

$$-\frac{d}{d\nu(x)}\left\{p(x)f'(x) - \int_a^x f(s)d\sigma(s)\right\} \tag{3.2.10}$$

would be similar to the one presented here since a change of variable (Appendix I) and use of Theorem 3.2.0 would reduce (3.2.10) to an expression of the form (3.2.2).

§3.3 THE WEYL CLASSIFICATION (cf., [81])

Let ν be right-continuous and locally of bounded variation on $[a, \infty)$. As we saw in section 3.1, ν induces a σ-finite signed Borel measure m_ν on $[a, \infty)$. If we let $|m_\nu|$ denote the total variation measure of ν then $|m_\nu|$ is equivalent [24, p. 126] to the measure m_V induced by the function

$$V(x) \equiv \int_a^x |d\nu(s)| \qquad x \in [a, \infty) \tag{3.3.0}$$

where the integral is a Lebesgue-Stieltjes integral, in general, but in our cases can be taken to be a Riemann-Stieltjes integral since the latter exists by the hypotheses imposed upon ν .

We define the Lebesgue space $L^2(|m_\nu| ; I)$ as the space of all equivalence classes of functions which are square-integrable with respect to the measure $|m_\nu|$. Two functions are equivalent if they are equal $[\nu]$.

Thus $f \in L^2(|m_\nu| ; I)$ if and only if

$$\int_I |f(x)|^2 d|m_\nu| < \infty . \tag{3.3.1}$$

The latter space is also equivalent to the space $L^2(V ; I)$ of functions f such that

$$\int_I |f(x)|^2 dV(x) < \infty \tag{3.3.2}$$

where V is defined in (3.3.0) and (3.3.2) is a Lebesgue-Stieltjes integral.

If $f \in L^2(V; I)$ we define the norm of f as

$$\|f\| \equiv \left\{ \int_I |f(x)|^2 dV(x) \right\}^{\frac{1}{2}} . \qquad (3.3.3)$$

The norm defined by (3.3.3) then turns $L^2(V; I)$ into a Hilbert space which we denote by H.

Let I be one of $[a, b]$ or $[a, \infty)$, $a > -\infty$, $b < \infty$. We shall use (3.2.2) to define an operator L on H. We let $\mathcal{D}_L \equiv \mathcal{D}$ be defined by

$$\mathcal{D} = \{f \in H : f \in \mathcal{D}_\nu(I) \text{ and } \ell[f] \in H\} . \qquad (3.3.4)$$

For $f \in \mathcal{D}$

$$Lf = \ell[f] \qquad (3.3.5)$$

where $\ell[f]$ is as in (3.2.2).

The domain $\mathcal{D}$ can be considered as the maximal linear manifold on which it is possible, by means of (3.3.5), to define an operator in H.

Some problems arise concerning the single-valuedness of the resulting operator L in H. In the case when ν is non-decreasing, sufficient conditions guaranteeing the single-valuedness of L can be found in [35, p. 260] and [36]. We shall assume hereafter that L defined by (3.3.5) is single-valued. (For the applications that we are dealing with, the operator L is single-valued.)

DEFINITION 3.3.1:

Let $I = [a, \infty)$. If for a particular $\lambda_0 \in \mathbb{C}$ every solution of the equation

$$Ly = \lambda_0 y \tag{3.3.6}$$

satisfies

$$\int_a^{\infty} |y(x)|^2 dV(x) < \infty \tag{3.3.7}$$

then L is said to be of limit-circle type at infinity. In the contrary case L is said to be of limit-point type at infinity.

We now proceed to show that the classification depends only on L and not on the λ chosen.

THEOREM 3.3.1:

If every solution of (3.3.6) is of class $L^2(V; I)$ for some $\lambda_0 \in \mathbb{C}$, then for any $\lambda \in \mathbb{C}$ every solution of $Ly = \lambda y$ is of class $L^2(V; I)$.

Proof: Let ϕ, ψ be linearly independent solutions of (3.3.6) satisfying

$$\phi(x)\psi'(x) - \phi'(x)\psi(x) = 1 . \tag{3.3.8}$$

Such solutions exist by Theorem 3.2.0 and [3, p. 348] (see

also [35, p. 220]). We remark that the $(')$ appearing in (3.3.8) is, in general, a right-derivative.

Let $\lambda \neq \lambda_0$ and y be a solution of

$$Ly = \lambda y \ .$$

Then

$$Ly = \lambda_0 y + (\lambda - \lambda_0) y \ . \tag{3.3.10}$$

We now use the variation of constants formula [3, p. 351] and apply it to (3.3.10) when it is rewritten in the equivalent form

$$\begin{aligned} y(x) &= y(a) + y'(a)(x-a) + \int_a^x (x-s)y(s)\,d\big(\sigma(s) - \lambda_0 \nu(s)\big) \\ &\quad -(\lambda - \lambda_0) \int_a^x (x-s)y(s)\,d\nu(s) \ , \end{aligned} \tag{3.3.11}$$

to obtain a representation of y as

$$\begin{aligned} y(x) &= \alpha\phi(x) + \beta\psi(x) \\ &\quad + (\lambda - \lambda_0) \int_c^x \{\phi(x)\psi(t) - \psi(x)\phi(t)\} y(t)\,d\nu(t) \end{aligned} \tag{3.3.12}$$

where α , β are constants and $c \leqq x$ is to be chosen later. If we write

$$\|y\|_c \equiv \left\{ \int_c^t |y(x)|^2 dV(x) \right\}^{\frac{1}{2}} \tag{3.3.13}$$

then since, $\phi, \psi \in L^2(V;I)$, it is possible to choose R such that

$$R \geqq \max\{\|\phi\|_c, \|\psi\|_c\} \tag{3.3.14}$$

for all $t \geqq c$.

Then

$$\left| \int_c^t \{\phi(x)\psi(t) - \psi(x)\phi(t)\}y(t)d\nu(t) \right|$$

$$\leqq \int_c^t |\phi(x)\psi(t)y(t)|dV(t) + \int_c^t |\psi(x)\phi(t)y(t)|dV(t) \tag{3.3.15}$$

and use of the Schwarz inequality gives

$$|\phi(x)| \int_c^t |\psi(t)y(t)|dV(t) \leqq |\phi(x)|\|\psi\|_c\|y\|_c$$

with a similar inequality holding for the other integral in (3.3.15).

Hence, by (3.3.14),

$$|\phi(x)| \int_c^t |\psi y|dV + |\psi(x)| \int_c^t |\phi y|dV \leqq (|\phi(x)| + |\psi(x)|)R\|y\|_c . \tag{3.3.16}$$

Consequently

$$\|y\|_c \leqq |\alpha|\|\phi\|_c + |\beta|\|\psi\|_c + |\lambda - \lambda_0|R\|y\|_c\{\|\phi\|_c + \|\psi\|_c\}$$

$$\leqq (|\alpha| + |\beta|)R + |\lambda - \lambda_0| \cdot 2R^2\|y\|_c .$$

We now choose c so large that, for all $t \geqq c$,

$$|\lambda - \lambda_0| R^2 < \frac{1}{4}$$

from which it will follow that

$$\|y\|_c \leqq 2R(|\alpha| + |\beta|)$$

for all $t \geqq c$. Thus letting $t \to \infty$ we find that $y \in L^2(V:(c, \infty))$ and hence is in $L^2(V; I)$.

It follows from this that in the limit-point case at most one linearly independent solution can be in $L^2(V; I)$.

<u>THEOREM 3.3.2:</u>

Let ν be non-decreasing and suppose that for $b > a$

$$\int_a^b |y(x, \lambda)|^2 \, d\nu(x) > 0 \qquad (3.3.17)$$

for all λ real or complex where $y \not\equiv 0$ is a solution of (3.3.9). If $\operatorname{Im} \lambda \neq 0$ then there exists at least one solution of (3.3.9) in $L^2(V; I)$.

<u>Proof</u>: This is the standard "nesting circles" analysis. Let ϕ, ψ be two linearly independent solutions of (3.3.9) satisfying

$$\phi(a, \lambda) = \sin \alpha \qquad \phi'(a, \lambda) = -\cos \alpha$$

$$\psi(a, \lambda) = \cos \alpha \qquad \psi'(a, \lambda) = \sin \alpha \qquad (3.3.18\text{-}19)$$

where $\alpha \in [0, \pi)$. Then ϕ , ψ , ϕ' , ψ' are entire functions of λ for fixed x (Appendix III, section 1). If we write

$$[\phi\bar{\psi}](x) \equiv \phi(x)\psi'(x) - \psi(x)\phi'(x) \tag{3.3.20}$$

then it follows from (3.3.18-19) that

$$[\phi\bar{\psi}](x) = 1 . \tag{3.3.21}$$

The solutions ϕ , ψ are real for real λ and satisfy

$$\phi(a, \lambda)\cos\alpha + \phi'(a, \lambda)\sin\alpha = 0$$

$$\psi(a, \lambda)\sin\alpha - \psi'(a, \lambda)\cos\alpha = 0 .$$

Every solution ϕ of (3.3.9) is, up to a constant multiple, of the form

$$\theta = \phi + m\psi \tag{3.3.22}$$

where m is some number which depends on λ . Now let $a < b < \infty$ and introduce a real boundary condition at b by requiring that

$$g(\lambda) \equiv y(b, \lambda)\cos\beta + y'(b, \lambda)\sin\beta = 0 \tag{3.3.23}$$

for $\beta \in [0, \pi)$. The eigenvalues of (3.3.9) are then the zeros of the entire function $g(\lambda)$. Since these eigenvalues must necessarily be real (Appendix III, Theorem III.1.2),

(3.3.23) does not vanish identically and consequently the zeros have no finite point of accumulation.

We now seek m such that the solution θ above satisfies the boundary condition (3.3.23). A simple computation shows that

$$m = -\frac{\cot \beta \phi(b , \lambda) + \phi'(b , \lambda)}{\cot \beta \psi(b , \lambda) + \psi'(b , \lambda)} .$$

Thus as λ , b , β vary $m = m(\lambda , b , \beta)$. Since $\phi , \phi' , \psi , \psi'$ are entire functions of λ (Appendix III, Theorem III.1.0), m is meromorphic in λ and real for real λ .

If we let $z = \cot \beta$ and fix b , λ for the moment, then as β varies from 0 to π , z varies over $(-\infty , \infty)$ so that the image of the real axis under the transformation

$$m = -\frac{Az + B}{Cz + D} , \tag{3.3.24}$$

where $A = \phi(b , \lambda)$, $B = \phi'(b , \lambda)$, $C = \psi(b , \lambda)$, $D = \psi'(b , \lambda)$ and $AD - BC \neq 0$, is a circle C_b in the m-plane. Thus θ will satisfy (3.3.23) if and only if m lies on C_b . From (3.3.24) we have

$$z = -\frac{B + Dm}{A + Cm} \tag{3.3.25}$$

so that the circle C_b , which is the image of $\operatorname{Im} z = 0$, is given by

$$(\bar{A} + \bar{C}\bar{m})(B + Dm) - (A + Cm)(\bar{B} + \bar{D}\bar{m}) = 0 \ . \qquad (3.3.26)$$

Since every circle with center γ and radius r can be described by

$$r^2 - |\gamma|^2 + \gamma\bar{\omega} + \bar{\gamma}\omega - \omega\bar{\omega} = 0 \qquad (3.3.27)$$

we see on comparing coefficients of (3.3.26-27) that center $\tilde{m}_b$ of C_b is given by

$$\tilde{m}_b = \frac{A\bar{D} - B\bar{C}}{\bar{C}D - \bar{D}C} \qquad (3.3.28)$$

and its radius r_b is given by

$$r_b = \frac{|AD - BC|}{|\bar{C}D - \bar{D}C|} \ . \qquad (3.3.29)$$

Substituting the values for A , B , C , D into (3.3.26) we obtain the equivalent equation

$$[\theta\theta](b) = 0 \ . \qquad (3.3.30)$$

In the same way we find that

$$[\phi\psi](b) = A\bar{D} - B\bar{C}$$

$$[\psi\psi](b) = \bar{D}C - \bar{C}D$$

$$1 = [\phi\bar{\psi}](b) = AD - BC \ .$$

Hence,

$$\tilde{m}_b = -\frac{[\phi\psi](b)}{[\psi\psi](b)} \qquad (3.3.31)$$

and

$$r_b = \frac{1}{|[\psi\psi](b)|} \ . \qquad (3.3.32)$$

The coefficient of $m\bar{m}$ in (3.3.26) is $[\psi\psi](b)$ and so the interior of C_b in the m-plane is given by

$$\frac{[\theta\theta](b)}{[\psi\psi](b)} < 0 \ . \qquad (3.3.33)$$

Since ψ , ψ' take on real values at a we have

$$\begin{aligned}[\psi\psi](b) - [\psi\psi](a) &= [\psi\psi](b) \\ &= \int_a^b (\bar{\psi}L\psi - \psi L\bar{\psi})\,d\nu\end{aligned}$$

by Theorem 3.2.1,

$$\begin{aligned}&= 2i\int_a^b \operatorname{Im}(\bar{\psi}L\psi)\,d\nu \\ &= 2i(\operatorname{Im}\lambda)\int_a^b |\psi|^2\,d\nu \ . \qquad (3.3.34)\end{aligned}$$

Similarly we can show

$$[\theta\theta](b) = [\theta\theta](a) + 2i(\operatorname{Im}\lambda)\int_a^b |\theta|^2\,d\nu \qquad (3.3.35)$$

where

$$[\theta\theta](a) = -2i\operatorname{Im}(m) \ . \qquad (3.3.36)$$

Combining (3.3.34 - 36) into (3.3.33) we obtain

$$\frac{\operatorname{Im}\lambda \int_a^b |\theta|^2 d\nu - \operatorname{Im}(m)}{\operatorname{Im}\lambda \int_a^b |\psi|^2 d\nu} < 0 \tag{3.3.37}$$

and so if we take it that $\operatorname{Im}\lambda > 0$, say, then

$$\int_a^b |\theta|^2 d\nu < \frac{\operatorname{Im}(m)}{\operatorname{Im}(\lambda)} . \tag{3.3.38}$$

The points m are on C_b if and only if

$$\int_a^b |\theta|^2 d\nu = \frac{\operatorname{Im}(m)}{\operatorname{Im}(\lambda)} . \tag{3.3.39}$$

The radius r_b of C_b is also given by

$$r_b = \frac{1}{2 \operatorname{Im}\lambda \int_a^b |\psi|^2 d\nu} . \tag{3.3.40}$$

If, say, ν is constant on some interval I then for $b \in I$ r_b = constant and so the circles remain the same until b lies outside I.

If $a < \gamma < b < \infty$, then

$$\int_a^\gamma |\theta|^2 d\nu \leqq \int_a^b |\theta|^2 d\nu < \frac{\operatorname{Im}(m)}{\operatorname{Im}(\lambda)}$$

and so

$$C_b \subseteq C_\gamma \quad \text{if} \quad b > \gamma\ . \tag{3.3.41}$$

Thus the sequence (C_b) of circles is "nested" in the sense (3.3.41).

Assuming that ν is not eventually constant the sequence (C_b) shall therefore converge to either a circle or a point as $b \to \infty$. If the C_b converge to a circle C_∞ then its radius which is necessarily given by the $\lim r_b$ in (3.3.40) is positive and hence

$$\int_a^b |\psi|^2 \, d\nu < \infty\ . \tag{3.3.42}$$

In this case if $\tilde{m}_\infty$ is any point on C_∞ then $\tilde{m}_\infty$ lies within C_b , $b > 0$.

Thus

$$\int_a^b |\phi + \tilde{m}_\infty \psi|^2 \, d\nu < \frac{\mathrm{Im}(\tilde{m}_\infty)}{\mathrm{Im}(\lambda)} \tag{3.3.43}$$

and so

$$\phi + \tilde{m}_\infty \psi \in L^2\big(V\ ;\ [a\ ,\infty)\big)$$

since we can let $b \to \infty$ in (3.3.43).

The same argument applies if C_∞ reduces to a point $\hat{m}_\infty$ since in this case

$$\phi + \hat{m}_\infty \psi \in L^2(V\,;\,[a\,,\infty)) \;. \qquad (3.3.44)$$

In the latter case θ is unique. Hence if $\mathrm{Im}(\lambda) \neq 0$ there is always a solution of (3.3.9) in L^2. The motivation for the terms "limit-circle" and "limit-point" is clearly expressed in the preceding discussion.

REMARK:

Difficulties arise if one only assumes that ν is of bounded variation in the preceding theorem. One such is that the eigenvalue problem (3.3.9), (3.3.23) may admit non-real eigenvalues (Chapter 4) so that one has to guarantee that the function $g(\lambda)$ does not vanish identically. (We shall show in the next chapter that, for differential and difference equations, the number of non-real eigenvalues must be finite. Thus, in these cases $g(\lambda)$ cannot vanish identically.)

Another difficulty arises because of the indefiniteness of the sign of

$$\int_a^b |y(x\,,\lambda)|^2 d\nu(x)$$

when y is a non-trivial solution of (3.3.9) and $b > a$: For varying b the latter integral may take on both positive and negative values thus making it difficult to interpret (3.3.37), say, for $b > a$.

In any case, even when ν is of bounded variation

(3.3.9) can have solutions in $L^2(V;I)$ for some λ (and hence for all λ by Theorem 3.3.1). Thus the "limit-point" and "limit-circle" cases can both occur for general ν. We shall see this in the following section.

§3.4 APPLICATIONS:

Let ν be a locally absolutely continuous function on $[a,\infty)$ such that $r(x) = \nu'(x) > 0$ a.e. Furthermore suppose that σ is locally absolutely continuous and $q(x) = \sigma'(x)$ a.e.

Then

$$\nu(x) = \nu(a) + \int_a^x r(s)\,ds \qquad (3.4.1)$$

and

$$\sigma(x) = \sigma(a) + \int_a^x q(s)\,ds\,. \qquad (3.4.2)$$

If we consider the problem

$$-\frac{d}{d\nu(x)}\left\{y'(x) - \int_a^x y\,d\sigma\right\} = \lambda y \qquad (3.4.3)$$

on $[a,\infty)$ then (3.4.3) reduces to

$$-\frac{d}{r(x)\,dx}\left\{y'(x) - \int_a^x yq\,ds\right\} = \lambda y \quad \text{a.e.}$$

and the latter is equivalent to

$$-y''(x) + q(x)y(x) = \lambda r(x)y(x) \tag{3.4.4}$$

on $[a, \infty)$. Thus (3.4.3) includes (3.4.4). A similar treatment allows one to include (3.4.4) when $r(x)$ is indefinite as to sign. Also of interest is that (3.4.3) is also defined when ν has intervals of constancy, e.g. when $r(x) = 0$ a.e. on some interval. A treatment of the Weyl classification for (3.4.4) when $r(x) > 0$ can be found in Hellwig [28].

Since (3.4.3) is equivalent to the integral equation

$$y(x) = \alpha + \beta(x-a) + \int_a^x (x-s)y(s)d\{\sigma(s) - \lambda v(s)\} \tag{3.4.5}$$

then using the methods outlined in Chapter 1 we can then include three-term recurrence relations in (3.4.3). For let (t_n) be a given sequence,

$$t_{-1} = a < t_0 < t_1 < \cdots < t_n < \cdots \to \infty$$

and let

$$t_n - t_{n-1} = \frac{1}{c_{n-1}} \qquad n = 0, 1, \ldots \tag{3.4.6}$$

where $c_{-1} > 0$.

We let σ be a right-continuous step-function with jumps at the points (t_n) where

$$\sigma(t_n) - \sigma(t_n - 0) = c_n + c_{n-1} - b_n \qquad n = 0, 1, \ldots \tag{3.4.7}$$

where (b_n) is a given sequence. Moreover by defining ν as a right-continuous step-function with jumps at the (t_n) where

$$\nu(t_n) - \nu(t_n - 0) = a_n \qquad (3.4.8)$$

where (a_n) is another given sequence then (3.4.5) will have a solution $y(t)$ whose values $y_n \equiv y(t_n)$ satisfy the recurrence relation

$$-c_n y_{n+1} - c_{n-1} y_{n-1} + b_n y_n = \lambda a_n y_n \ . \qquad (3.4.9)$$

We note here that (3.4.6) implies

$$\sum_{0}^{\infty} \frac{1}{c_{n-1}} = \infty \qquad (3.4.10)$$

though this requirement can be omitted by the introduction of a new function $p(t)$ which is right-continuous and defined by (1.1.24) with $p(t) > 0$. We can then consider the more general

$$-\frac{d}{d\nu(x)} \left\{ p(x) y'(x) - \int_a^x y \, d\sigma \right\} = \lambda y \qquad (3.4.11)$$

which would then reduce to (3.4.9) without (3.4.10).

The construction in section 3.3 also applies in the case (3.4.9). For example the radius of the circle C_b is given by (3.3.40) thus if we let

$$a = t_{-1} < t_0 < t_1 < \cdots < t_{m-1} = b$$

then

$$\int_a^b |\psi|^2 d\nu = \sum_0^{m-1} a_n |\psi_n|^2 , \tag{3.4.12}$$

(since we always assume that ν is continuous at a).

Consequently

$$r_b = \frac{1}{2(\operatorname{Im} \lambda) \sum_0^{m-1} a_n |\psi_n|^2} . \tag{3.4.13}$$

For the latter result see [3, pp. 125-26 and equation (5.4.6)]. The nesting circle analysis for three-term recurrence relations can be found in [3, pp. 125-29].

Moreover the space $L^2(V ; I)$ is then equivalent to the space of square-summable sequences "with respect to the weight a_n ", i.e. $(y_n) \in \ell^2(V ; I)$ if and only if

$$\sum_0^{\infty} |a_n y_n^2| < \infty .$$

Thus if $c_n > 0$, b_n is any sequence and $a_n \geqq 0$, then [3, p. 129, Theorem 5.4.2] for any λ $\operatorname{Im} \lambda \neq 0$,

$$-c_n y_{n+1} - c_{n-1} y_{n-1} + b_n y_n = \lambda a_n y_n$$

has at least one nontrivial solution $\omega = (\omega_n)$ in $\ell^2(V ; I)$.

§3.5 LIMIT-POINT AND LIMIT-CIRCLE CRITERIA:

In this section we shall give some conditions on ν and σ which will enable us to establish the limit-point or limit-circle classification of

$$-\frac{d}{d\nu(x)}\left\{y'(x) - \int_a^x y\,d\sigma\right\} = \lambda y \qquad (3.5.0)$$

in the space $L^2(V\,;\,I)$ where $V(x)$ is the total variation of $\nu(x)$ over the interval $[a\,,\,x]$ and $I = [a\,,\,\infty)$, $a > -\infty$. This space was defined in (3.3.3).

LEMMA 3.5.0:

Let $\sigma\,,\,\nu$ be right-continuous functions locally of bounded variation on $[a\,,\,\infty)$. Suppose that there exists $\lambda_0 \in \mathbb{R}$ such that $\sigma(x) - \lambda_0\nu(x)$ is non-decreasing for $x \geqq a$, say.

Then (3.5.0), with $\lambda = \lambda_0$, has a solution $y(x)$ with

$$y(x) \geqq 1 \qquad x \geqq a\,. \qquad (3.5.1)$$

Proof: Let $y(x)$ be the solution of (3.5.0) satisfying the initial conditions

$$y(a) = 1 \quad , \quad y'(a) = 0\,. \qquad (3.5.2)$$

Then, by Theorem 3.2.0, $y(x)$ is a solution of the integral equation

$$y(x) = 1 + \int_a^x (x-s)y(s)d\big(\sigma(s) - \lambda_0\nu(s)\big) \qquad (3.5.3)$$

for $x \geqq a$. Since $y(a) = 1$ then by continuity there exists an interval $[a, a+\delta]$, $\delta > 0$ in which $y(x) > 0$. Then, for $x \in [a, a+\delta]$ the integral in (3.5.3) is non-negative and so

$$y(x) \geqq 1 \qquad x \in [a, a+\delta] . \qquad (3.5.4)$$

Since $y(a+\delta) \geqq 1$ there exists $\delta_1 > 0$ such that $y(x) > 0$ in $[a+\delta, \delta_1]$. Consequently for x in such an interval

$$y(x) = y(a+\delta) + \int_{a+\delta}^x (x-s)y(s)d\big(\sigma(s) - \lambda_0\nu(s)\big)$$

and so

$$y(x) \geqq 1 \qquad x \in [a+\delta, \delta_1] .$$

Repeating this process we obtain an increasing sequence (δ_n) of real numbers. It is then necessary that

$$\lim_{n\to\infty} \delta_n = \infty \qquad (3.5.5)$$

otherwise if $\lim \delta_n = \delta^*$ then $y(\delta^*) \geqq 1$ by continuity and so we could repeat the above process past δ^*. This contradiction proves that (3.5.5) holds and thus

$$y(x) \geqq 1 \qquad x \geqq a \ .$$

<u>THEOREM 3.5.1</u>: Let σ, ν satisfy all the hypotheses of Lemma 3.5.0. Suppose further that

$$\int_a^\infty |d\nu(t)| = \infty \ . \tag{3.5.6}$$

Then (3.5.0) is limit-point at ∞ .

<u>Proof</u>: It suffices to show that, for some λ , there is a solution of (3.5.0) which is not in $L^2(V\,;\,I)$. Let $\lambda = \lambda_0$ where the latter exists by hypothesis. From Lemma 3.5.0 there exists a solution $y(t)$ of (3.5.0) such that (3.5.1) holds.

Then for such a solution,

$$\int^\infty |y(x)|^2 |d\nu(t)| \geqq \int_a^\infty |d\nu(t)| = \infty$$

hence y is not in $L^2(V\,;\,I)$.

<u>COROLLARY 3.5.1</u>:

Let (a_n) be a sequence such that

$$\sum_0^\infty |a_n| = \infty \ .$$

Let (b_n) be any given sequence and (c_n) another positive sequence.

If there exists a real number λ_0 such that

$$b_n - c_n - c_{n-1} + \lambda_0 a_n > 0 \qquad n = 0, 1, \ldots \tag{3.5.7}$$

then

$$c_n y_{n+1} + c_{n-1} y_{n-1} - b_n y_n = \lambda a_n y_n \tag{3.5.8}$$

is limit-point at ∞, i.e. for some λ there corresponds a solution (y_n) such that

$$\sum_0^\infty |a_n||y_n|^2 = \infty . \tag{3.5.9}$$

Proof: We note here in passing that Lemma 3.5.0 extends to equations of the form (3.4.11) when $p(x) > 0$ right-continuous and of bounded variation satisfying $p(t)^{-1}$ locally $L(a, \infty)$. The proof is similar with minor changes.

We define a step-function $\nu(t)$ with jumps at the (t_n) by

$$\nu(t_n) - \nu(t_n - 0) = -a_n \tag{3.5.10}$$

and require that ν be constant on $[t_{n-1}, t_n)$, $n = 0, 1, \ldots$. We define $\sigma(t)$ as in (3.4.7). Then (3.5.0) has solutions $y(t)$ such that $y(t_n) = y_n$ satisfies the recurrence relation (3.5.8).

(3.5.10) and the hypothesis imply that (3.5.6) is satisfied. Moreover for $\lambda = \lambda_0$, (3.5.7) implies that

$\sigma - \lambda_0 \nu$ is non-decreasing. Thus Theorem 3.5.1 applies and so

$$\int_a^\infty |y(t)|^2 \, |d\nu(t)| = \infty$$

which implies (3.5.9)

In this form, Corollary 3.5.1 is a minor extension of [3, p. 135, Theorem 5.8.2] where the case $a_n > 0$ is considered.

THEOREM 3.5.2:

Let σ, ν be right-continuous functions locally of bounded variation on $[0, \infty)$ and

$$\int_0^\infty t\,|d\sigma(t)| < \infty . \tag{3.5.11}$$

Then a necessary and sufficient condition for (3.5.0) to be limit-circle is that

$$\int_0^\infty t^2\,|d\nu(t)| < \infty . \tag{3.5.12}$$

Proof: We rewrite the solution of (3.5.0) as

$$y(x) = \alpha + \beta x + \int_0^x (x-s)\,y(s)\,d\big(\sigma(s) - \lambda\nu(s)\big) . \tag{3.5.13}$$

Then (3.5.0) is limit-circle if say (3.5.13) with $\lambda = 0$ has only $L^2(V;I)$ solutions. Using now a result in [3, p. 389, Theorem 12.5.2] we find that (3.5.13) with $\lambda = 0$ has a pair of solutions y, z such that

$$y(x) \to 1 \qquad x \to \infty \tag{3.5.14}$$

$$z(x) \sim x \qquad x \to \infty\,. \tag{3.5.15}$$

The solutions are then linearly independent. If (3.5.0) is limit circle then these solutions must belong to $L^2(V\,;\,I)$. Since $z(x) \sim x$ for large x we will have

$$\int_X^\infty |z(s)|^2\,|d\nu(s)| = \int_X^\infty \left|\frac{z(s)}{s}\right|^2 s^2\,|d\nu(s)|$$

$$> c\int_X^\infty s^2\,|d\nu(s)|\,. \tag{3.5.16}$$

Hence (3.5.12) is satisfied. A similar calculation shows that since $y \to 1$ then this forces ν to be of bounded variation over $[0\,,\infty)$.

Next, if (3.5.11-12) are satisfied then ν must be of bounded variation over $[0\,,\infty)$ and hence

$$\int_0^\infty |y(x)|^2\,|d\nu(x)| < \infty$$

on account of (3.5.14). Since $z(x) \sim x$ an argument similar to the one leading to (3.5.16) shows that

$$\int_0^\infty |z(x)|^2\,|d\nu(x)| = O\left\{\int_0^\infty t^2\,|d\nu(t)|\right\}$$

and thus z is square-integrable. Consequently Theorem 3.3.1 implies that (3.5.0) is limit circle. This completes the proof.

If we take it that $\sigma(t) \equiv$ constant on $[0, \infty)$ then (3.5.11) is trivially satisfied and thus we find that

$$-\frac{d\,y'(x)}{d\nu(x)} = \lambda y \qquad (3.5.17)$$

is limit-circle at infinity if and only if (3.5.12) holds. The latter result extends a theorem of M.G. Krein [39, p. 882, Theorem 1] who proved Theorem 3.5.2 for equations of the form (3.5.17) and ν non-decreasing.

We note here that (3.5.11) is not superfluous, i.e. the negation of (3.5.11) does not, in general, produce the conclusion of the theorem.

For let

$$\nu(t) = \int_0^t e^{-x}\,dx + 1 = -e^{-t}$$

$$\sigma(t) = \int_0^t 1\,dx = t .$$

Then (3.5.0) is equivalent to

$$-y'' + y = \lambda e^{-t} y \qquad t \in [0, \infty) . \qquad (3.5.18)$$

Here

$$\int_0^\infty t\,|d\sigma(t)| = \infty$$

while

$$\int_0^\infty t^2 |d\nu(t)| < \infty .$$

However (3.5.18) is limit-point since for $\lambda = 0$ it has the solution $y(t) = \exp t$ which is not in $L^2(V : I)$ since ν is non-decreasing. Thus (3.5.11) cannot be omitted and as we shall see presently, it is not sufficient that σ be of bounded variation over $[0, \infty)$. In fact, if we only assume that

$$\int_0^\infty t^\delta |d\sigma(t)| < \infty \qquad 0 \leqq \delta < 1 \tag{3.5.19}$$

then (3.5.12) is no longer both necessary and sufficient for (3.5.0) to be limit circle.

For let

$$\nu(t) = \int_0^t (x+1)^{-4}\, dx$$

$$\sigma(t) = \frac{3}{4} \int_0^t (x+1)^{-2}\, dx .$$

Then (3.5.0) reduces to

$$-y'' + \frac{3}{4(x+1)^2}\, y = \lambda \frac{y}{(x+1)^4} \quad \text{on} \quad [0, \infty) . \tag{3.5.20}$$

A computation shows that (3.5.12) is satisfied along with (3.5.19). However (3.5.20) is limit-point since for $\lambda = 0$ it admits the solution

$$y(t) = (t+1)^{3/2}$$

which is not in $L^2(V;I)$.

<u>COROLLARY 3.5.2</u>:

Let b_n , c_n be real sequences $c_n > 0$ for all n and such that

$$\sum_{n=0}^{\infty} \left\{ \sum_{0}^{n} \frac{1}{c_{j-1}} \right\} |c_n + c_{n-1} - b_n| < \infty . \qquad (3.5.21)$$

Then a necessary and sufficient condition that (3.5.8) be in the limit-circle case is that

$$\sum_{n=0}^{\infty} \left\{ \sum_{j=0}^{n} \frac{1}{c_{j-1}} \right\}^2 |a_n| < \infty . \qquad (3.5.22)$$

<u>Proof</u>: We define ν , σ as in Corollary 3.5.1. Then (3.5.21) implies (3.5.11) and (3.5.22) is equivalent to (3.5.12). The result follows after we note that

$$t_n = \sum_{0}^{n} \frac{1}{c_{j-1}} \qquad n = 0 , 1 , \ldots$$

since $t_{-1} = 0$.

Results similar to those of Corollary 3.5.1-2 can be stated for differential equations of the form

$$-y'' + q(t)y = \lambda r(t)y \qquad t \in [0 , \infty)$$

by defining ν , σ as the indefinite integrals of r , q

respectively and then interpreting (3.5.11-12) in this case.

§3.6 J-SELF-ADJOINTNESS OF GENERALIZED DIFFERENTIAL OPERATORS:

In the sequel we shall be interpreting the notion of limit-point, associated with (3.5.0), in terms of a property of the domain of an operator defined by this expression. It will turn out that the concept of limit-point is equivalent to the "J-self-adjointness" of a particular operator in a Krein space (Appendix III.3). When the function ν in (3.5.0) is non-decreasing and ν, σ are absolutely continuous, the above-mentioned equivalence is well known, since we are now dealing with a second order differential operator in a Hilbert space. See for example [46, §17 and §18.3]. In the course of the proof of the more general equivalence, we shall be adapting an argument of Everitt [15, p. 42] to generalized differential expressions.

In the following we shall assume that ν, σ are right-continuous functions both locally of bounded variation on $I = [a, \infty)$, $a > -\infty$. As usual we denote the total variation of ν by V. The operator L generated by the "formally self-adjoint" generalized differential expression

$$\ell[y](x) = -\frac{d}{d\nu(x)}\left\{ y'(x) - \int_a^x y(s)\,d\sigma(s)\right\} \qquad (3.6.0)$$

is defined as in section 3.2 of this chapter. Thus the domain $\mathcal{D}$ of L consists of all functions $f \in L^2(V; I)$ such that

i) f is locally absolutely continuous on I.

ii) f has at each point $x \in [a, \infty)$ a right-derivative $f'_+(x) \equiv f'(x)$.

iii) The function

$$\mu(x) \equiv f'(x) - \int_a^x f(s)d\sigma(s)$$

is V-absolutely continuous locally on I .

iv) $\ell[f](x) \in L^2(V; I)$.

For $f \in \mathcal{D}$ L is defined by

$$Lf = \ell[f] . \qquad (3.6.1)$$

The notions of "regularity" and "complete regularity" of the expression (3.6.0) are defined in [35, p. 249] in the case when ν is non-decreasing.

In general we shall say that the end a is regular if the set of "points of growth" of $V(x)$ and the set of its values is bounded from below and if the set of points of growth of σ is bounded below and, in addition σ is of bounded variation in some right-neighborhood of a . If a is not regular then it is said to be singular. The end a is completely regular if a belongs to the interval concerned. These definitions are due to Kac [35].

In our case $I = [a, \infty)$. It is then clear from the latter definitions and the basic assumptions on ν, σ of section 3.2 that the end a is completely regular. If

$I = [a, b]$ then the ends are both completely regular.

We note that since ν, σ are continuous at a, b, in the case of a finite interval, the left and right-derivatives of a solution $y(t)$ of (3.5.0) will exist there and be equal. This can also be seen by extending ν, σ past a, b by setting each equal to $\nu(a), \sigma(a)$ and $\nu(b), \sigma(b)$ respectively on some interval containing $[a, b]$.

THEOREM 3.6.1:

Let $I = [a, b]$ be a finite interval and consider $\ell[\cdot]$ on I. Let g be any function in $L^2(V, I)$. The equation

$$\ell[y] = g \tag{3.6.2}$$

has a solution $y(x)$ satisfying

$$y(a) = y(b) = 0 \tag{3.6.3}$$

$$y'(a) = y'(b) = 0 \tag{3.6.4}$$

if and only if the function $g(x)$ is J-orthogonal to all solutions of the homogeneous equation $\ell[y] = 0$.

Proof: We note that f is J-orthogonal to g if and only if

$$\int_a^b f(x)\bar{g}(x)d\nu(x) = 0 . \tag{3.6.5}$$

(The J-orthogonality stems from the J-inner product in $L^2(V; I)$, see Appendix III.3.)

This theorem can be proved exactly as in [46, p. 62, Lemma 1]. For by Theorem 3.2.0, and Theorem I.3.1 the equation (3.6.2) has a unique solution which satisfies $y(a) = 0$, $y'(a) = 0$.

Let z_1, z_2 be a fundamental system of solutions of $\ell[z] = 0$ which satisfy

$$z_1(b) = 1 \quad , \quad z_2(b) = 0$$

$$z_1'(b) = 0 \quad , \quad z_2'(b) = 1 \; .$$

Applying Theorem 3.2.1 to y and z_j we find

$$\int_a^b g(x)\bar{z}_j(x)\,d\nu(x) = \int_a^b \ell[y](x)\bar{z}_j(x)\,d\nu(x)$$

$$= \int_a^b y(x)\,\overline{\ell[z_j](x)}\,d\nu(x) - [y'\bar{z}_j - y\bar{z}_j']_a^b \, . \qquad (3.6.6)$$

By noting that $\ell[z_j] = 0$, $j = 1, 2$, and using the boundary conditions above, (3.6.6) reduces to

$$\int_a^b g(x)\bar{z}_j(x)\,d\nu(x) = \begin{cases} -y'(b) & j = 1 \\ y(b) & j = 2 \end{cases} \, . \qquad (3.6.7)$$

Thus (3.6.4) is satisfied if and only if (3.6.7) vanishes for

$j = 1, 2$, i.e. f if J-orthogonal to a fundamental system of solutions of $\ell[z] = 0$ and thus the conclusion follows.

Now since the measure induced by ν, in (3.6.0), is absolutely continuous with respect to the measure induced by V the quantity (section 3.1)

$$\frac{d\nu(x)}{dV(x)} \quad \text{exists} \quad [V] . \tag{3.6.8}$$

Consequently, the expression

$$-\frac{d}{dV(x)}\left\{y'(x) - \int_a^x y\,d\sigma\right\} = -\frac{d\nu(x)}{dV(x)} \cdot \frac{d}{d\nu(x)}\left\{y'(x) - \int_a^x y\,d\sigma\right\} \tag{3.6.9}$$

V-almost everywhere by [24, p. 135, Ex. 1, and Theorem A].

Thus if we denote by $\ell^{\dagger}[y]$ the expression defined by $y \in \mathcal{D}_{\nu}$ (section 3.2),

$$\ell^{\dagger}[y](x) \equiv -\frac{d}{dV(x)}\left\{y'(x) - \int_a^x y\,d\sigma\right\} \tag{3.6.10}$$

we see that $\ell^{\dagger}$ is another generalized differential expression related to ℓ by (3.6.9), i.e. for $y \in \mathcal{D}_{\nu}$

$$\ell^{\dagger}[y](x) = \frac{d\nu}{dV}(x) \cdot \ell[y](x) \tag{3.6.11}$$

and both of these are defined on the same domain $\mathcal{D}_{\nu}$. Thus (3.6.10) gives rise to an operator $L^{\dagger}$ on $\mathcal{D}$, where D is the domain of L defined earlier, such that for $y \in D$,

$$L^{\dagger}y = \frac{d\nu}{dV} \cdot Ly \qquad (3.6.12)$$

or, in terms of the Gram operator J defined in Appendix III.3,

$$L^{\dagger} = JL \ . \qquad (3.6.13)$$

If, in Theorem 3.6.1, we assume that ν is non-decreasing then we can replace $\ell[y]$ in (3.6.2) by $\ell^{\dagger}[y]$ and the conclusion will then follow with J-orthogonality being the usual orthogonality in $L^2(\nu; I)$, since $\nu \equiv V$ in this case.

We now define a new operator, denoted by L_0', with domain $\mathcal{D}_0'$ defined by [46, p. 60]

$$\mathcal{D}_0' \equiv \{f \in \mathcal{D} : f \equiv 0 \text{ outside a finite interval } [\alpha, \beta] \subset (a, b)\} \qquad (3.6.14)$$

The restriction of the operator L to $\mathcal{D}_0'$ defines L_0'. Thus for $y \in \mathcal{D}_0'$

$$L_0'y = Ly = \ell[y] \ . \qquad (3.6.15)$$

Similarly we can define $(L_0')^{\dagger}$ by $y \in (\mathcal{D}_0')^{\dagger} = \mathcal{D}_0'$

$$(L_0') = L^{\dagger}y = \ell^{\dagger}[y] \ . \qquad (3.6.16)$$

<u>THEOREM 3.6.2</u>:

a) If $y \in \mathcal{D}_0'$, $z \in \mathcal{D}$ then

$$[L_0' y , z] = [y , Lz] \qquad (3.6.17)$$

where $[\ ,\ ]$ is the J-inner product defined by the left hand side of (3.6.5).

Moreover, the operator L_0' is J-hermitian, i.e.

$$[L_0' y , z] = [y , L_0' z] \qquad y , z \in \mathcal{D}_0' . \qquad (3.6.18)$$

b) If $y \in \mathcal{D}_0'$, $z \in \mathcal{D}$ then writing $L_1 \equiv (L_0')^\dagger$ we have

$$(L_1 y , z) = (y , L^\dagger z)$$

where $(\ ,\ )$ is the inner product in $L^2(V ; I)$.

Again, the operator L_1 is hermitian, i.e.

$$(L_1 y , z) = (y , L_1 z) \qquad y , z \in \mathcal{D} . \qquad (3.6.19)$$

<u>Proof</u>: Both a), b) can be shown as in [46, p. 61] making use of Theorem 3.2.1 so we omit the details.

We now proceed as in [46, §17] in defining the operators L_0 and $L_0^\dagger$

Suppose that the interval $[a , b]$ is finite. (Then $\ell , \ell^\dagger$ are both regular on $[a , b]$.)

We define the domain $\mathcal{D}_0$ of the operator L_0 by

$$\mathcal{D}_0 = \{y \in \mathcal{D} : y(a) = y(b) = y'(a) = y'(b) = 0\} \qquad (3.6.20)$$

and, for $y \in \mathcal{D}_0$,

$$L_0 y = Ly \qquad (3.6.21)$$

$$L_0^{\dagger} y = L^{\dagger} y . \qquad (3.6.22)$$

THEOREM 3.6.3:

For any $y \in \mathcal{D}_0$, $z \in \mathcal{D}$

$$[L_0 y , z] = [y , Lz] \qquad (3.6.23)$$

$$(L_0^{\dagger} y , z) = (y , L^{\dagger} z) \qquad (3.6.24)$$

and the operator L_0 is J-hermitian while $L_0^{\dagger}$ is hermitian, i.e. for any $y , z \in \mathcal{D}_0$,

$$[L_0 y , z] = [y , L_0 z] \qquad (3.6.25)$$

$$(L_0^{\dagger} y , z) = (y , L_0^{\dagger} z) . \qquad (3.6.26)$$

Proof: We refer to [46, p. 62, I, II] since the proofs are similar.

LEMMA 3.6.1:

Let $R_0^\dagger \equiv$ range of $L_0^\dagger$ and let M be the set of all solutions of the equation $\ell^\dagger[z] = 0$.

Then

$$H = L^2(V\,;\,I) = R_0^\dagger + M\,. \tag{3.6.27}$$

Proof: Since all solutions of the homogeneous equation are continuous functions on $[a, b]$, they all belong to H and so $M \subset H$. It is also readily seen that M is a finite dimensional subspace in H of dimension 2 . If y is a solution of (3.6.2), with ℓ replaced by $\ell^\dagger$, then y is in D_0 . Hence $\ell^\dagger[y] = L_0^\dagger y$. Thus the existence of y implies that $g \in R_0^\dagger$. Theorem 3.6.1 then states that g lies in $R_0^\dagger$ if and only if it is orthogonal to M . Since H is a Hilbert space the decomposition (3.6.27) follows.

THEOREM 3.6.4:

The domain D_0 of the operator L_0 is dense in H .

Proof: Since the domain D_0 is the same for the operators L_0 and $L_0^\dagger$ it suffices to show that every element h orthogonal to D_0 is zero. Letting h be such an element, we have then $(h, y) = 0$ for all $y \in D_0$. Let z be any solution of $\ell^\dagger[z] = h$. For $y \in D_0$ we have, by Theorem 3.6.3,

$$(z\,,\,L_0^{\dagger}y) = (L^{\dagger}z\,,\,y) = (\ell^{\dagger}[z]\,,\,y) = (h\,,\,y) = 0$$

and so z is orthogonal to $R_0^{\dagger}$. Consequently $z \in M$ by the previous lemma, and so $\ell^{\dagger}[z] = 0$, i.e. $h = 0$.

We recall that a set is dense in a Krein space if it is dense in the Hilbert norm topology. Thus the latter theorem expresses the fact that the domain $\mathcal{D}_0$ of the operator L_0 is dense in the Krein space H.

THEOREM 3.6.5:

The operators L_0 and $L_0^{\dagger}$ are J-symmetric and symmetric respectively.

Proof: This follows from Theorems 3.6.3 - 4.

Note: The rest of the results in [46, §17] can be similarly shown to be true in this more general setting. Thus, e.g., in the regular case the operator L_0 is a closed J-symmetric operator whose J-adjoint L_0^{x} is equal to L [46, p. 66, Theorem 1].

In the singular case, i.e. when $I = [a\,,\,\infty)$, $a > -\infty$, we follow the approach outlined in [46, §17.4] where we begin with the operator L_0' defined in (3.6.14-15). We recall that L_0' and L_1 are both defined on the same domain $\mathcal{D}_0'$.

THEOREM 3.6.6:

The domain of definition $\mathcal{D}_0'$ of L_0' is dense in H and L_0' is therefore a J-symmetric operator.

Proof: An argument similar to that of [46, p. 68] shows that $\mathcal{D}_0'$, when viewed as the domain of L_1, is dense in H. Thus L_1 is a symmetric operator, by Theorem 3.6.2, and L_0' is J-symmetric.

We now take the closure of L_0', $\tilde{L}_0'$ in the Hilbert space topology and define

$$L_0 = \tilde{L}_0'$$

it then follows from the preceding theorem that L_0 is a closed J-symmetric operator.

We now proceed to find a property of the domain $\mathcal{D}$ of $L^{\dagger}$ (and so L) when $\ell^{\dagger}$ is in the limit-point case in $L^2(V\,;\,I)$.

LEMMA 3.6.2: [14].

For any set of six functions $\{f_p : 1 \leqq p \leqq 3\}$ $\{g_q : 1 \leqq q \leqq 3\}$ each being locally absolutely continuous on $[0, \infty)$ and each having a finite right-derivative at each point $x \in [0, \infty)$,

$$\det\{[f_p q_q](x)\} = 0 \qquad x \in [0, \infty)$$

where

$$[fg](x) \equiv f(x)\bar{g}_+'(x) - \bar{g}(x)f_+'(x) .$$

<u>Proof</u>: See [14, p. 374].

<u>LEMMA 3.6.3</u>:

Let μ be a Borel measure on $[0, \infty)$ and let $L^2(\mu)$ be the space of square integrable "functions" with respect to μ. Suppose that f, g are complex-valued μ-measurable functions which satisfy

$$f \in L^2(\mu ; [0, \infty)) \quad , \quad g \in L^2(\mu ; [0, X))$$

for all $X > 0$ and that

$$g \notin L^2(\mu ; [0, \infty)) .$$

Then

$$\lim_{X\to\infty} \left| \int_0^X f\bar{g}\, d\mu \right| \left\{ \int_0^X |g|^2\, d\mu \right\}^{-\frac{1}{2}} = 0 .$$

<u>Proof</u>: This result can be proven in exactly the same way as in [15, p. 42] with the necessary modifications.

Let $L^\dagger$ be the operator defined by (3.6.12), (3.6.10). Then by Theorem 3.3.2, for $\operatorname{Im} \lambda \neq 0$, the problem

$$\ell^{\dagger}[y] = \lambda y \quad \text{on} \quad [0\,,\infty) \tag{3.6.28}$$

has at least one nontrivial solution in $L^2(V\,;I)$ where $I = [0\,,\infty)$. Using this result and Theorem 3.3.1 along with Lemmas 3.6.1-2 we can show, by adapting the argument of Everitt [15, pp. 42 - 45] to our situation, that whenever (3.6.28) is limit-point there follows

$$\lim_{x\to\infty} \{f(x)\bar{g}'_+(x) - f'_+(x)\bar{g}(x)\} = 0 \qquad f\,,g \in \mathcal{D} \tag{3.6.29}$$

Conversely if (3.6.28) is limit-circle, then it must be so for $\lambda = 0$. In this case it is possible to find two real linearly independent solutions $\phi\,,\psi$ of

$$\ell^{\dagger}[y] = 0 \tag{3.6.30}$$

which satisfy (3.3.18-19) say. By hypothesis these are in $L^2(V\,;I)$ and consequently in D . Moreover, by (3.3.18-19)

$$\phi(x)\psi'_+(x) - \phi'_+(x)\psi(x) = 1$$

hence

$$\lim_{x\to\infty} [\phi\psi](x) \neq 0\ .$$

Summarizing, we obtain

<u>THEOREM 3.6.7</u>:

A necessary and sufficient condition for (3.6.28) to

be limit-point is that for all $f, g \in \mathcal{D}$, (3.6.29) be satisfied.

THEOREM 3.6.8:

Let σ satisfy (3.5.11). Then (3.6.28) is limit-point if and only if

$$\ell[y] = \lambda y \tag{3.6.31}$$

is limit-point (in the $L^2(V;I)$ sense) where ℓ and $\ell^\dagger$ are related by (3.6.11).

Proof: For Theorem 3.5.2 implies that (3.6.28) is limit-circle if and only if

$$\int_0^\infty t^2 dV(t) = \infty \tag{3.6.32}$$

where $V(t) = \int_0^t |d\nu(s)|$. However the latter is equivalent to

$$\int_0^\infty t^2 |d\nu(t)| = \infty \tag{3.6.33}$$

and Theorem 3.5.2 again implies that (3.6.31) is limit circle if and only if (3.6.33) and so (3.6.32) is satisfied. The result now follows.

COROLLARY 3.6.1:

In order that (3.6.31) be in the limit-point case at infinity (in the space $L^2(V;I)$) it is necessary and sufficient that

$$\lim_{x\to\infty} [fg](x) = 0 \qquad f , g \in \mathcal{D} \tag{3.6.34}$$

where [ ] is defined in Lemma 3.6.2 and $\mathcal{D}$ is the domain of L , $L^{\dagger}$ defined earlier.

We now define the operators L_{α} , $L_{\alpha}^{\dagger}$. Let the domain $\mathcal{D}_{\alpha}$ of L_{α} be defined by

$$\mathcal{D}_{\alpha} = \{f \in \mathcal{D} : \ f(0)\cos\alpha - f_{+}'(0)\sin\alpha = 0\} \tag{3.6.35}$$

where $\alpha \in [0 , \pi)$, and for $f \in \mathcal{D}_{\alpha}$,

$$L_{\alpha} f = Lf \ . \tag{3.6.36}$$

Similarly $L_{\alpha}^{\dagger}$ is defined on the same domain $\mathcal{D}_{\alpha}$ and

$$L_{\alpha}^{\dagger} f = L^{\dagger} f \qquad f \in \mathcal{D}_{\alpha} \tag{3.6.37}$$

or what is the same, $L_{\alpha}^{\dagger} = J L_{\alpha}$. We now proceed to show that, in the limit-point case, $L_{\alpha}^{\dagger}$ is self-adjoint.

First of all we note that if $f , g \in \mathcal{D}_{\alpha}$ then

$$[fg](0) = 0 \ . \tag{3.6.38}$$

Next the Lagrange identity (Theorem 3.2.1) shows that, for $f , g \in \mathcal{D}_{\alpha}$, $X > 0$

$$\int_0^X \{f(x)\overline{\ell^{\dagger}[g](x)} - \ell[f](x)\overline{g(x)}\}dV(x)$$

$$= -[fg](x) + [fg](0) . \qquad (3.6.39)$$

Consequently if $\ell^{\dagger}$ is limit-point and $f, g \in \mathcal{D}_{\alpha}$, we let $X \to \infty$ in (3.6.39) and use Theorem 3.6.7 and (3.6.38) to find

$$(L_{\alpha}^{\dagger}f, g) = (f, L_{\alpha}^{\dagger}g) \qquad f, g \in \mathcal{D}_{\alpha} \qquad (3.6.40)$$

and so $L_{\alpha}^{\dagger}$ is symmetric ($\mathcal{D}_{\alpha}$ is dense in $L^2(V; I)$ since it contains the domain $\mathcal{D}_0$ of the operator L_0 defined in the singular case by $L_0 = \tilde{L}_0'$. The proof of this is similar to that in [46, p. 71, VI]). In the same fashion it can be shown that

$$[L_{\alpha}f, g] = [f, L_{\alpha}g] \qquad f, g \in \mathcal{D}_{\alpha} \qquad (3.6.41)$$

so that L_{α} is J-symmetric.

<u>THEOREM 3.6.9</u>: In the limit-point case, the domain $\mathcal{D}_{\alpha}$ is a domain of self-adjointness of $L_{\alpha}^{\dagger}$ if and only if $\mathcal{D}_{\alpha}$ has the following properties,

i) For all $f, g \in \mathcal{D}_{\alpha}$, $[fg](0) = 0$,

ii) If $g \in \mathcal{D}$ satisfies $[fg](0) = 0$ for all $f \in \mathcal{D}_{\alpha}$, then $g \in \mathcal{D}_{\alpha}$.

Proof:

We note that this result is a particular case of a theorem of Naimark [46, p. 73, Theorem 1] and can be proven similarly.

With $\mathcal{D}_\alpha$ defined as in (3.6.35) a simple computation shows that both (i) and (ii) are satisfied in Theorem 3.6.9 and consequently, in the limit-point case, $L_\alpha^\dagger$ is self-adjoint. On the other hand, if $L_\alpha^\dagger$ is self-adjoint then the deficiency indices [46, p. 26] of the operator are $(0\ ,\ 0)$. Consequently the equation

$$(L_\alpha^\dagger)^* z = \lambda z \qquad (3.6.41)$$

has no non-trivial solution in $L^2(V\ ;\ I)$. Since $L_\alpha^\dagger$ is self-adjoint (3.6.41) implies that the problem

$$L^\dagger z = \lambda z$$

$$z(0)\cos\alpha - z'(0)\sin\alpha = 0$$

has no solutions in $L^2(V\ ;\ I)$. Thus (3.6.28) is limit-point. Hence we have proved

THEOREM 3.6.10:

The equation (3.6.28) is in the limit-point case in $L^2(V\ ;\ I)$ if and only if the operator $L_\alpha^\dagger$, $\alpha \in [0\ ,\ \pi)$, is self-adjoint.

In the following discussion $(L_\alpha^\dagger)^*$, $L_\alpha^\times$, will denote the Hilbert space adjoint and Krein space adjoint of the operators $L_\alpha^\dagger$, L_α respectively. Let us suppose that $L_\alpha^\dagger$ is self-adjoint. Then by Theorem 3.6.10, (3.6.28) is in the limit point case and so L_α is a J-symmetric operator and its J-adjoint $L_\alpha^\times$ exists. We denote its domain by $\mathcal{D}_\alpha^\times$.

Let $f \in \mathcal{D}_\alpha$, $g \in \mathcal{D}_\alpha^\times$

$$[L_\alpha f , g] = [f , L_\alpha^\times g]$$
$$= (f , JL_\alpha^\times g) \ . \qquad (3.6.43)$$

Now since $L_\alpha^\dagger = JL_\alpha$, $L_\alpha^\dagger = (L_\alpha^\dagger)^* = (JL_\alpha)^* \supset L_\alpha^* J$. Moreover

$$JL_\alpha^\times = L_\alpha^* J \ . \qquad (3.6.44)$$

Substituting (3.6.44) into (3.6.43) we find that $g \in \mathcal{D}(L_\alpha^* J)$. But

$$\mathcal{D}(L_\alpha^* J) \subset \mathcal{D}(L_\alpha^\dagger) \equiv \mathcal{D}_\alpha$$

hence $g \in \mathcal{D}_\alpha$. Consequently $\mathcal{D}_\alpha^\times \subset \mathcal{D}_\alpha$ and since L_α is J-symmetric $\mathcal{D}_\alpha \subset \mathcal{D}_\alpha^\times$. Hence $\mathcal{D}_\alpha = \mathcal{D}_\alpha^\times$ and so L_α is J-self-adjoint.

On the other hand, let L_α be J-self-adjoint for each $\alpha \in [0 , \pi)$, so that we have $L_\alpha = L_\alpha^\times$ or

$$[f, L_\alpha g] = [L_\alpha f, g] \qquad f, g \in \mathcal{D}_\alpha . \tag{3.6.45}$$

Using the Lagrange identity in (3.6.45) we find that for $f, g \in \mathcal{D}_\alpha$

$$0 = \lim_{X\to\infty} \int_0^X \{f\overline{L_\alpha g} - \bar{g}L_\alpha f\}d\nu = \lim_{X\to\infty} [\bar{g}f' - \bar{g}'f]_0^X .$$

Since $f, g \in \mathcal{D}_\alpha$, $\bar{g}(0)f'(0) - \bar{g}'(0)f(0) = 0$ so that

$$\lim_{x\to\infty} \{f(x)\bar{g}'(x) - f'(x)\bar{g}(x)\} = 0 \tag{3.6.46}$$

for all $f, g \in \mathcal{D}_\alpha$. We now wish to show that the latter equality in fact holds for all $f, g \in \mathcal{D}$. For this it would suffice to show that if f, g are any two real-valued functions, $f \in \mathcal{D}_\alpha$, $g \in \mathcal{D}_\beta$, $\alpha, \beta \in [0, \pi)$ then (3.6.46) holds. For if $f, g \in \mathcal{D}$

$$[fg](x) \equiv f(x)\bar{g}'(x) - f'(x)\bar{g}(x)$$

$$= [f_R g_R](x) + [f_I g_I](x) + i\{[f_I g_R](x) + [g_I f_R](x)\}$$

where $f = f_R + if_I$, $g = g_R + ig_I$. Hence for given $f \in \mathcal{D}$, $f_R \in \mathcal{D}_\alpha$ for some $\alpha \in [0, \pi)$ since $f_R(0)$, $f_R'(0)$ are real. A similar result holds for $f_I \in \mathcal{D}_\beta$ where $\beta \neq \alpha$, $\beta \in [0, \pi)$. The result now follows.

Thus let f, g be two real valued functions with $f \in \mathcal{D}_\alpha$, $g \in \mathcal{D}_\beta$, $\beta \neq \alpha$. We will show that under certain hypotheses on σ, ν we can find a function $g^* \in \mathcal{D}_\alpha$ such that

$$[fg](x) = [fg^*](x) \qquad \text{for all large } x .$$

One such condition is the following: Let σ be ν-absolutely continuous so that for some ν-measurable function ϕ,

$$d\sigma = \phi d\nu$$

in the sense of the measures defined by σ and ν. Let $g^*(x)$ be defined by

$$g^*(x) = \begin{cases} g(x) , & x \geq 1 \\ ax + b , & 0 \leqq x < 1 \end{cases} \tag{3.6.47}$$

where a, b are to be determined. We need $g^*(x)$ to be absolutely continuous so that

$$g^*(1) = g(1) . \tag{3.6.48}$$

We also wish to have $g^* \in \mathcal{D}_\alpha$ where α is as above. Consequently

$$g^*(0)\cos\alpha - g^{*\prime}(0)\sin\alpha = 0 \tag{3.6.49}$$

(3.6.48-49) then determine a, b in (3.6.47). The resulting function $g^*(x)$ has the following properties:

$g^* \in AC_{loc}(0, \infty)$ and $g_+^{*\prime}(x)$ exists for each $x \geqq 0$.

$$G(x) \equiv g_+^{*\prime}(x) - \int_0^x g^*(s)\,d\sigma(s) \qquad (3.6.50)$$

is then defined for $x \geqq 0$ and in fact

$$G(x) = a - \int_0^x g^*(s)\phi(s)\,d\nu(s) \qquad (3.6.51)$$

for $x \geqq 0$. Since $g^* = g$ for $x \geqq 1$, $G(x)$ is ν-absolutely continuous if $x \geqq 1$, since $g \in \mathcal{D}_\beta$. Furthermore (3.6.51) implies that $G(x)$ is ν-absolutely continuous if $0 \leqq x < 1$. Hence $G(x)$ is ν-absolutely continuous and

$$Lg^* \in L^2(V; [0, \infty))$$

since Lg is. Finally (3.6.49) implies that $g^* \in \mathcal{D}_\alpha$. Thus given $g \in \mathcal{D}_\beta$, there exists $g^* \in \mathcal{D}_\alpha$ such that $g^* \equiv g$ for large x and therefore for $f \in \mathcal{D}_\alpha$,

$$\lim_{x\to\infty} [fg](x) = \lim_{x\to\infty} [fg^*](x)$$

$$= 0$$

by (3.6.46). From the preceding discussion it then follows that

$$[fg](x) = O(1) \qquad f, g \in \mathcal{D}.$$

Theorem 3.6.7 now implies that (3.6.28) is in the limit-point case and Theorem 3.6.10 then forces $L_\alpha^\dagger$, $\alpha \in [0, \pi)$ to be self-adjoint. Hence the J-self-adjointness of L_α implies the self-adjointness of $L_\alpha^\dagger$ under the above-mentioned restrictions on σ, ν. When σ, ν are both step-functions with jumps at the same points then the absolute continuity of σ with respect to ν is satisfied and thus the resulting difference operator is J-self-adjoint in the Krein space if and only if it is self-adjoint. A similar result holds for ordinary differential operators since in this case both σ, ν are absolutely continuous in the ordinary sense. In fact by weakening this hypothesis on σ, ν we may include more general operators other than those previously mentioned.

<u>THEOREM 3.6.11</u>:

Suppose

$$\int_0^\infty t|d\sigma(t)| < \infty . \tag{3.6.52}$$

Then a necessary and sufficient condition for the operator L_α to be J-self-adjoint is that

$$\int_0^\infty t^2|d\nu(t)| = \infty . \tag{3.6.53}$$

<u>Proof</u>: For the assumption (3.6.52) implies that (3.6.28) is limit-point if and only if (3.6.53) holds, by Theorem 3.5.2. The latter, being limit-point, is necessary and sufficient

for the operator $L_\alpha^\dagger$ to be self-adjoint which is equivalent to L_α being J-self-adjoint from the preceding discussion.

The above theorem is a minor extension of a result quoted by Langer [4, p. 122], in the particular case of these generalized differential expressions with $\sigma(t) = \text{const}$. As we saw in section 3.5 the requirement (3.6.52) cannot be waived, in general. Using the methods of Chapters 1-2 it is possible to give the discrete analogs of all the above theorems, however this task will not be undertaken here for the sake of brevity. For the relation between the notion of limit-point and the existence of spectral functions, in the case when ν is non-decreasing, we refer to [35], [38] and the bibliographies therein. In particular, where σ is constant and ν is non-decreasing the result in Theorem 3.6.11 is equivalent to the existence of a unique spectral function [38, p. 75]. Thus, a similar result applies to difference equations by means of the usual construction (cf.,[64],[65],[66])

REMARKS:

1. In Theorem 3.6.8 the hypothesis that σ should satisfy (3.5.11) can be omitted without affecting the conclusion. This follows essentially from (3.6.11). Since if (3.6.28) is limit-circle then it must be so for $\lambda = 0$. Thus it can only have solutions in $L^2(V; I)$ for such a λ . However, (3.6.11) states that ℓ and $\ell^\dagger$ have the same

solutions to either homogeneous equation. Thus $\ell^\dagger$ is limit-circle if and only if ℓ is, by Theorem 3.3.1. Consequently the result follows.

2. The latter discussion therefore implies (Theorem 3.6.7) that (3.6.31) is limit-point if and only if (3.6.29) holds. Hence the notion of limit-point for (3.6.31) is equivalent to the existence of a related J-self-adjoint operator. For problems related to the expansion into eigenfunctions of the problem (3.6.31), with a homogeneous boundary condition at the origin, we cite the paper of Daho and Langer [11]. The expansion and completeness theorems of generalized differential operators were also treated, when ν is assumed non-decreasing, in a paper of Kac [37]. In these arguments it is necessary to assume, in the limit-point case, that the resolvent set of the J-self-adjoint operators be non-empty. In the next chapter we will show that, in the regular case, the resolvent sets of second-order difference and differential operators with "indefinite weight functions" are non-empty. In the singular case it is not known whether or not the resolvent set of the J-self-adjoint operator generated by an ordinary Sturm-Liouville problem with indefinite weight function need be empty or not (see the footnote in [11, p. 171]). It would seem that the spectrum of such an operator necessarily consists of at most a finite number of non-real eigenvalues because of the result

in the regular case.

§3.7 DIRICHLET INTEGRALS ASSOCIATED WITH GENERALIZED DIFFERENTIAL EXPRESSIONS:

In this section we examine various properties of the "maximal domain" $\mathcal{D}$ defined in the earlier section. The basic material for this section can be found in [17]. We shall extend the various Notions contained therein to cover generalized differential expressions and thus, in particular, three-term recurrence relations.

DEFINTION 3.7.1:

i) The operator L, defined in (3.3.5), with domain $\mathcal{D}$ is said to have the *Dirichlet property* (DI) *at infinity* if

$$\int_a^\infty |f_+'(x)|^2 dx < \infty , \qquad f \in \mathcal{D} \tag{3.7.1}$$

and

$$\int_a^\infty |f(x)|^2 |d\sigma(x)| < \infty , \qquad f \in \mathcal{D} . \tag{3.7.2}$$

ii) L is *Conditionally Dirichlet* (CD) *at infinity* if

$$\lim_{x\to\infty} \int_a^x f\bar{g}\, d\sigma \qquad \text{exists} \tag{3.7.3}$$

and is finite for all $f , g \in \mathcal{D}$ and if (3.7.1) holds for all $f \in \mathcal{D}$.

iii) L is said to be *Strong Limit-Point* (SLP) at infinity if

$$\lim_{x\to\infty} f(x)\bar{g}_+'(x) = 0 \qquad f, g \in \mathcal{D} . \tag{3.7.4}$$

iv) L is *Limit-Point* (LP) at infinity if

$$\lim_{x\to\infty} \{f(x)\bar{g}_x'(x) - f_+'(x)\bar{g}(x)\} = 0 \qquad f, g \in \mathcal{D} . \tag{3.7.5}$$

This definition is consistent with the usual definition of section 3.6 because of Theorem 3.6.7 and Remark 1.

It follows immediately from (iii) - (iv) that if L is strong limit point then L is limit-point, i.e.

$$\text{SLP} \Rightarrow \text{LP} . \tag{3.7.6}$$

The converse is not valid in general (an example will be given later). Similarly it follows from the definitions that if L is Dirichlet then it is conditionally Dirichlet, i.e.

$$\text{DI} \Rightarrow \text{CD} \tag{3.7.7}$$

with the converse false in general [17, p. 313]. The implication CD => SLP was shown to be valid by Everitt [17, pp. 313-14] for ordinary differential expressions. In the general case the proof is not very different.

For suppose that L is CD at infinity. We find

upon integrating by parts that, for all $f, g \in \mathcal{D}$,

$$\int_a^x f(\overline{Lg})d\nu = (f\bar{g}')(a) - (f\bar{g}')(x) + \int_a^x (f'\bar{g}' + f\bar{g}d\sigma) .$$

Since L is CD the right-hand integral tends to a finite limit as $x \to \infty$. Moreover since $f, Lg \in L^2(V; I)$ the left-hand integral tends to a finite limit as $x \to \infty$. Consequently, for every $f, g \in \mathcal{D}$,

$$\alpha \equiv \lim_{x\to\infty} (f\bar{g}')(x) \quad \text{exists and is finite.} \tag{3.7.8}$$

If possible, let us assume that $\alpha \neq 0$ for some $f, g \in \mathcal{D}$. Then

$$\lim_{x\to\infty} |f(x)\bar{g}'(x)| = |\alpha| > 0 .$$

Thus for $x \geqq X$,

$$|f(x)\bar{g}'(x)| > \frac{1}{2}|\alpha| . \tag{3.7.9}$$

If f is uniformly bounded above on $[X, \infty)$ then the latter inequality implies that $g'(x)$ is uniformly bounded away from zero. Consequently $g' \notin L^2(X, \infty)$ which is a contradiction.

If f is not uniformly bounded then there exists an increasing sequence $\{x_n\}$ with $x_n \to \infty$ along which

$$f(x_n) \to \infty \quad , \quad n \to \infty \; .$$

(3.7.9) then implies that $|f(x)| > 0$ for $x \geqq X_1$, say, and hence

$$|f'(x)\bar{g}'(x)| > \frac{1}{2}\,|\alpha|\,\left|\frac{f'(x)}{f(x)}\right| \qquad x \geqq X_1 \; .$$

Integrating the latter over $[X_1 , x_n]$ and letting $n \to \infty$ we find

$$\int_{X_1}^{\infty} |f'(x)\bar{g}'(x)|dx = \infty$$

a contradiction, by the Schwarz inequality, since both $f' , g' \in L^2(a , \infty)$. Hence the conclusion is that CD => SLP . Again, in general, this implication is irreversible [17, p. 313]. Thus

$$\text{DI} \Rightarrow \text{CD} \Rightarrow \text{SLP} \Rightarrow \text{LP} \; . \qquad (3.7.10)$$

We now interpret these results for three-term recurrence relations, the theory having been developed in the case of ordinary differential expressions.

§3.8 DIRICHLET CONDITIONS FOR THREE-TERM RECURRENCE RELATIONS:

Let (c_n) , (b_n) be real sequences and suppose that $c_n > 0$ for all n . Let (a_n) be a sequence of real numbers where $a_n \neq 0$ for all n. Let (t_n) be an increasing sequence of real numbers defined by

$$t_n - t_{n-1} = \frac{1}{c_{n-1}} \qquad n = 0, 1, \ldots \tag{3.8.1}$$

where $c_{-1} > 0$ and $t_{-1} = a$ is fixed. We also assume that $t_n \to \infty$ as $n \to \infty$.

Now define step-functions ν, σ by requiring that both ν, σ be constant on $[t_{n-1}, t_n)$, $n = 0, 1, \ldots$ and that these have discontinuities at the (t_n) only, given by

$$\sigma(t_n) - \sigma(t_n - 0) = c_n + c_{n-1} - b_n, \qquad n = 0, 1, \ldots$$

and

$$\nu(t_n) - \nu(t_n - 0) = -a_n, \qquad n = 0, 1, \ldots. \tag{3.8.2-3}$$

We also suppose that ν, σ are both continuous at a and that neither have a jump at infinity.

Let ϕ be summable and consider the differential equation

$$\ell[y](x) = -\frac{d}{d\nu(x)}\left\{y'(x) - \int_a^x y\, d\sigma\right\} = \phi(x) \qquad x \in [a, \infty) \tag{3.8.4}$$

where ν, σ are defined above. Rewriting the solution of the above as the solution of a Volterra-Stieltjes integral equation we see that, using the methods of Chapter 1, the solution $y(t)$ is linear on $[t_{n-1}, t_n)$ and if $y_n \equiv y(t_n)$ then y_n satisfies the recurrence relation

$$-c_n y_{n+1} - c_{n-1} y_{n-1} + b_n y_n = a_n \phi_n \qquad n = 0, 1, \ldots \tag{3.8.5}$$

where $\phi_n \equiv \phi(t_n)$.

Thus the domain $\mathcal{D}$ of the operator L generated by the generalized differential expression above consists of polygonal curves, i.e. each function in $\mathcal{D}$ is absolutely continuous and linear on $[t_{n-1}, t_n)$ for $n = 0, 1, \ldots$. Moreover the space $L^2(V; I)$ becomes, in this case, the space $\ell^2(|a|)$, i.e. $f \in \ell^2(|a|)$ if

$$\sum_0^\infty |a_n||f_n|^2 < \infty \tag{3.8.6}$$

where $f_n \equiv f(t_n)$.

Since the domain $\mathcal{D}$ is essentially

$$\mathcal{D} = \{f \in L^2(V; I) : \ell[f](x) \in L^2(V; I)\} \tag{3.8.7}$$

we see therefore that a function $f \in \mathcal{D}$ if and only if the sequence (f_n) satisfies

$$\sum_0^\infty |a_n||f_n|^2 < \infty$$

and if

$$\ell[f]_n \equiv \frac{-c_n f_{n+1} - c_{n-1} f_{n-1} + b_n f_n}{a_n} \qquad n = 0, 1, \ldots \tag{3.8.8}$$

then

$$\ell[f]_n \in \ell^2(|a|) .$$

Hence the resultIng "difference operator" has domain $\mathcal{D}$ given by

$$\mathcal{D} = \{f = (f_n) \in \ell^2(|a|) : \ell[f]_n \in \ell^2(|a|)\} \tag{3.8.9}$$

where $\ell[f]_n$ is defined in (3.8.8). We have to keep in mind that each such sequence defines a function which is linear on $[t_{n-1}, t_n)$ and belongs to the domain $\mathcal{D}$ defined by (3.8.7). Thus we identify the domain (3.8.7) with (3.8.9). So if $f \in \mathcal{D}$ then $f'(t)$ is constant on $[t_{n-1}, t_n)$ and

$$f'(t) = \frac{f_n - f_{n-1}}{t_n - t_{n-1}}$$

$$= c_{n-1}(f_n - f_{n-1}) \qquad t \in [t_{n-1}, t_n) \tag{3.8.10}$$

Thus if $f \in \mathcal{D}$,

$$\int_a^\infty |f_+'(x)|^2 dx = \sum_0^\infty \int_{t_{n-1}}^{t_n} |f_+'(x)|^2 dx$$

$$= \sum_0^\infty c_{n-1}^2 |f_n - f_{n-1}|^2 \cdot (t_n - t_{n-1})$$

$$= \sum_0^\infty c_{n-1} |\Delta f_{n-1}|^2 \ . \tag{3.8.11}$$

Similarly for $f \in \mathcal{D}$,

$$\int_a^\infty |f(x)|^2 |d\sigma(x)| = \sum_0^\infty \int_{t_{n-1}}^{t_n} |f(x)|^2 |d\sigma(x)|$$

$$= \sum_0^\infty \int_{t_n-0}^{t_n+0} |f(x)|^2 |d\sigma(x)|$$

since σ is constant on $[t_{n-1}, t_n)$, $n = 0, 1, \ldots$.

$$= \sum_0^\infty |f_n|^2 |c_n + c_{n-1} - b_n| \ . \tag{3.8.12}$$

<u>DEFINITION 3.8.1:</u>

The difference operator L defined on (3.8.9) by

$$Lf = \ell[f] \tag{3.8.13}$$

where

$$\ell[f] = \ell[f]_n = a_n^{-1}\{b_n f_n - c_{n-1} f_{n-1} - c_n f_{n+1}\} \tag{3.8.14}$$

is said to have the *Dirichlet property at infinity* if for all $f = (f_n)$, $g = (g_n)$ in $\mathcal{D}$ we have

$$\sum_0^\infty c_{n-1} |\Delta f_{n-1}|^2 < \infty \tag{3.8.15}$$

and

$$\sum_{0}^{\infty} |c_n + c_{n-1} - b_n||f_n|^2 < \infty . \qquad (3.8.16)$$

Next let $x \in [t_n , t_{n+1})$. Then for $f , g \in \mathcal{D}$,

$$\int_a^x f(t)\bar{g}(t)d\sigma(t) = \sum_0^n \int_{t_j-0}^{t_j+0} f\bar{g}d\sigma + \int_{t_n+0}^{x} f\bar{g}\,d\sigma$$

$$= \sum_0^n f_j \bar{g}_j (c_j + c_{j-1} - b_j)$$

since σ is constant on (t_n , x) . This motivates

DEFINITION 3.8.2:

The difference operator L is *Conditionally Dirichlet at infinity* if for all $f , g \in \mathcal{D}$, (3.8.15) holds and

$$\lim_{m\to\infty} \sum_0^m f_n \bar{g}_n (c_n + c_{n-1} - b_n) \qquad (3.8.17)$$

exists and is finite.

Let $f , g \in \mathcal{D}$. Then

$$\lim_{x\to\infty} \{f(x)\bar{g}_+'(x) - f_+'(x)\bar{g}(x)\} \qquad (3.8.18)$$

$$= \lim_{n\to\infty} \{f(t_n)\bar{g}_+'(t_n) - f_+'(t_n)\bar{g}(t_n)\}$$

$$= \lim_{n\to\infty} \{f_n c_n \overline{\Delta g_n} - \bar{g}_n c_n \Delta f_n\}$$

by (3.8.10).

$$= \lim_{n\to\infty} c_n \{f_n \bar{g}_{n+1} - f_{n+1} \bar{g}_n\} \qquad (3.8.19)$$

whenever either of (3.8.18-19) exists. From the latter also stems the relation

$$\lim_{x\to\infty} f(x)\bar{g}'(x) = \lim_{n\to\infty} c_n f_n \overline{\Delta g_n} \qquad f, g \in \mathcal{D}. \qquad (3.8.20)$$

DEFINITION 3.8.3:

The difference operator L is said to be in the *Strong Limit-Point* case at infinity if for all $f, g \in \mathcal{D}$

$$\lim_{n\to\infty} c_n f_n \overline{\Delta g_n} \quad \text{exists} \quad (= 0). \qquad (3.8.21)$$

L is said to be in the *Limit-Point* case at infinity if for all $f, g \in \mathcal{D}$

$$\lim_{n\to\infty} c_n \{f_n \bar{g}_{n+1} - f_{n+1} \bar{g}_n\} = 0. \qquad (3.8.22)$$

The latter is consistent with the usual definition of limit-point for a three-term recurrence relation. (See for example, [3, pp. 498-99], [32, p. 425, Theorem 2].)

We note, in passing, that the theory developed in section 3.6 also includes the difference operators as special cases. Thus (3.8.22) holds for all $f, g \in \mathcal{D}$ if and only if a certain difference operator L_α defined by

$$L_\alpha f = \ell[f] \qquad f \in \mathcal{D} \tag{3.8.23}$$

and

$$f_{-1}\cos\alpha - c_{-1}(f_0 - f_{-1})\sin\alpha = 0 \tag{3.8.24}$$

is J-self-adjoint in the Krein space $\ell^2(|a|)$. If $a_n > 0$ for all n , then L_α is self-adjoint and consequently every symmetric extension of L_α must coincide with L_α . (This statement is also true in the Kreĭn space setting.) When (3.8.22) is satisfied it implies the self-adjointness of the "maximal" operator L [3, p. 499], it then follows that $L_\alpha = L$.

Moreover the implications in (3.7.10) are valid and generally irreversible.

EXAMPLE 3.8.1:

Let $c_n = n$, $a_n = 1$ and let

$$y_n = \begin{cases} \dfrac{1}{\sqrt{n}} & \text{if } n = 2^m \text{ some } m \geqq 0 \\ \dfrac{1}{n} & \text{if } n \neq 2^m \end{cases} .$$

Define b_n by

$$b_n = \frac{c_n y_{n+1} + c_{n-1} y_{n-1}}{y_n} \qquad n = 1, 2, \ldots$$

where $y_0 = 0$ say. A computation shows that $y_n \in \ell^2$ and if z_n is a linearly independent solution then we must have

$$y_n z_{n+1} - y_{n+1} z_n = \frac{\text{const}}{n} \qquad n = 1, 2, \ldots$$

by the discrete analog of the Wronskian identity. Thus if $z_n \in \ell^2$ the Schwarz inequality applied to the latter identity would produce a contradiction since the left side would be finite while the right side diverges. Thus

$$c_n y_{n+1} + c_{n-1} y_{n-1} - b_n y_n = 0 \qquad (3.8.25)$$

is LP . However

$$\lim_{n\to\infty} n\, y_n \,\Delta y_n$$

does not even exist. In fact,

$$\liminf_{n\to\infty} n\, y_n \,\Delta y_n = -1 \quad , \quad \limsup_{n\to\infty} n\, y_n \,\Delta y_n = 0 \ .$$

Thus (3.8.25) is not SLP. Other examples may be found to show that, in general, the implications (3.7.10) are irreversible even for three-term recurrence relations. The next result follows from remarks 1, 2 of the preceding section.

<u>THEOREM 3.8.1</u>:

A necessary and sufficient condition for

$$-c_n y_{n+1} - c_{n-1} y_{n-1} + b_n y_n = \lambda a_n y_n \qquad (3.8.26)$$

to be LP in the $\ell^2(|a|)$-sense is that (3.8.22) should hold for all $f, g \in \mathcal{D}$ (here $a_n \neq 0$) .

COROLLARY 3.8.1:

Let $|a_n| > \delta > 0$ for all n and that $0 < c_n < M$ for all n .

Then (3.8.26) is limit-point in the $\ell^2(|a|)$-sense independently of the coefficient b_n .

Proof: If we let $f \in \mathcal{D}$, then

$$\sum |f_n|^2 < \delta^{-1} \sum |a_n||f_n|^2 < \infty .$$

Thus $f_n \to 0$ as $n \to \infty$ for every $f \in \mathcal{D}$. Since the c_n are bounded (3.8.22) holds for all $f, g \in \mathcal{D}$. The result now follows.

If we let $c_n = a_n = 1$ in (3.8.26) we find that the equation

$$\Delta^2 y_{n-1} + b_n y_n = \lambda y_n \qquad n = 0, 1, \ldots$$

is always LP in the ℓ^2-sense (see [32, p. 436] and [3, p. 499]).

Unlike the results in Chapter 2, the limit-point,

limit-circle theory of difference equations differs substantially from the analogous theory for differential equations. One reason for this appears to be related to the general limit-point criterion (3.6.29) and its interpretation (3.8.22) for recurrence relations. For if we set $c_n = a_n = 1$ and let $f = (f_n) \in \ell^2$ then it is necessary that $\lim f_n$ should exist and be zero. Consequently (3.8.22) is automatically satisfied. On the other hand if we consider the differential equation

$$y'' + b(x)y = \lambda y \qquad x \in [0, \infty)$$

then the maximal domain of the operator generated by the expression is a subset of $L^2(0, \infty)$. Thus if $f \in L^2$, f need not tend to a limit at infinity and can be essentially unbounded. Hence (3.6.29) is far from being satisfied and thus conditions have to be imposed upon $b(x)$ to ensure that, say, if $f \in \mathcal{D}$ then f and f' have limits at infinity and (3.6.29) be satisfied.

In the following theorem we show that it is possible to strengthen the conclusion of Corollary 3.8.1 under the same set of hypotheses.

THEOREM 3.8.2:

Let $a_n \neq 0$, and c_n satisfy the hypotheses of Corollary 3.8.1.

Then

$$(Ly)_n \equiv a_n^{-1}\{-c_n y_{n+1} - c_{n-1} y_{n-1} + b_n y_n\} \qquad n = 0, 1, \ldots$$

(3.8.27)

has the Dirichlet property at infinity.

Proof: According to Definition 3.8.1 it is necessary to show (3.8.15-16).

Let $f \in \ell^2(|a|)$ be such that $(Lf)_n \in \ell^2(|a|)$. Since $|a_n| > \delta > 0$ then $f \in \ell^2$, i.e. the sequence f_n is square-summable in the usual sense. Since $f = (f_n)$ is in ℓ^2 then

$$\Delta f = (\Delta f_n) = (f_{n+1} - f_n) \in \ell^2 ,$$

and the hypothesis $c_n < M$ implies then that

$$\sum_0^\infty c_{n-1}|\Delta f_n|^2 < M \sum_0^\infty |\Delta f_n|^2 < \infty \quad \text{for} \quad f \in \mathcal{D} .$$

The same argument shows that

$$\sum |c_n + c_{n-1}||f_n|^2 < \infty \qquad f \in \mathcal{D} . \tag{3.8.28}$$

It is now sufficient to show that

$$\sum_0^\infty |b_n||f_n|^2 < \infty \qquad f \in \mathcal{D} \tag{3.8.29}$$

since (3.8.28-29) both will imply (3.8.16). To this end multiply (3.8.27) by $a_n \bar{y}_n$. Then

$$b_n|y_n|^2 = a_n(Ly)_n\bar{y}_n + c_n y_{n+1}\bar{y}_n + c_{n-1}y_{n-1}\bar{y}_n$$

Thus for $y \in \mathcal{D}$,

$$\sum_0^\infty |b_n||y_n|^2 < \sum_0^\infty |(Ly)_n \bar{y}_n||a_n + M\sum_0^\infty |y_{n+1}||\bar{y}_n|$$

$$+ M\sum_0^\infty |y_{n-1}||\bar{y}_n| .$$

Since y is ℓ^2 the last two series on the right are finite by the Schwarz inequality. Moreover since $y, Ly \in \ell^2(|a|)$ then

$$\sum_0^\infty |(Ly)_n \bar{y}_n||a_n| \leq \left\{\sum_0^\infty |a_n||(Ly)_n|^2\right\}^{\frac{1}{2}} \cdot \left\{\sum_0^\infty |a_n||y_n|^2\right\}^{\frac{1}{2}}$$

by the Schwarz inequality again. The series on the right being finite it now follows that

$$\sum_0^\infty |b_n||y_n|^2 < \infty$$

and so the conclusion follows.

COROLLARY 3.8.2:

Set $c_n = 1$, $a_n = -1$ in (3.8.27) and replace b_n by $b_n + 2$. Then the operator

$$(Ly)_n = \Delta^2 y_{n-1} + b_n y_n \qquad (3.8.30)$$

is in the Dirichlet condition at infinity independently of b_n . In particular (3.8.30) is always limit-point at infinity. (cf., also [75]).

CHAPTER 4

INTRODUCTION:

The study of second-order Sturm-Liouville problems with an indefinite weight-function dates back to the turn of the century. During the past ten years or so, it has been a topic of current research. We shall not attempt to give a detailed history of the subject here though we shall mention some aspects of the theory, especially those dealing with eigenvalues and their distribution along the real axis and in the complex plane. We shall be dealing exclusively with the history of differential equations since nothing appears to be known about three-term recurrence relations with indefinite weight-functions; though we shall prove a theorem in section 4.1 dealing with this topic. The problem

$$-y'' + q(x)y = \lambda k(x)y \tag{4.0.0}$$

$$y(0) = y(1) = 0 \tag{4.0.1}$$

where $q(x) \geqq 0$ and $k(x)$ has both signs in $[0, 1]$ was considered by Hilbert [29] who proved in this case that (4.0.0-1) admits an infinite sequence of eigenvalues

$$\cdots \lambda_{-2} < \lambda_{-1} < 0 < \lambda_0 < \lambda_1 < \cdots \qquad (4.0.2)$$

with no finite limit point and such that the eigenfunctions $y_i(x)$, $y_{-i}(x)$ vanish (i-1)-times in the interval (0 , 1) . This result was subsequently extended, to more general boundary conditions, by Mason [43] though the oscillation (c.f., [67 theorem was shown only when $k(x) \geqq 0$: For arbitrary $k(x)$ the result was proven by Picone [48] and simultaneously by Sanlievici [56] and Richardson [51]. The general case, that is when both q and k are indefinite, along with the boundary conditions (4.0.1), was considered by Richardson [52]. In the latter paper Richardson shows that the oscillation theorem for the eigenfunctions is valid for sufficiently large values of $|i|$. This was followed by another paper, [53], which essentially gave a survey of second order problems with an indefinite weight-function. In this same paper a more detailed proof of the oscillation theorem in the indefinite case was given, thus raising the question as to what happens to the "missing oscillation numbers". In answer to this, it seems that these are related to the non-real eigenvalues of the problem though his attempt at proving their existence appears dubious.* Still, the following interesting theorem on the oscillation properties of non-real eigenfunctions corresponding to non-real eigenvalues is proved.

---

* See the note in Appendix III, p. 307.

<u>THEOREM A</u>: [53, p. 302, Theorem X].

Let $q(x)$, in (4.0.0), be negative in a subinterval of the interval $[0, 1]$ and suppose that $k(x)$ changes sign once only in $[0, 1]$. Let λ be a non-real eigenvalue of (4.0.0-1) and let $y(x, \lambda) = u(x, \lambda) + iv(x, \lambda)$ be the corresponding eigenfunction.

Then the zeros of the real and imaginary parts, $u(x, \lambda)$, $v(x, \lambda)$, separate one another.

In this section we shall complement the results of [53] by finding an upper bound for the number of non-real eigenvalues of (4.0.0-1) in the indefinite case. It turns out that the analogous theorem for a three-term recurrence relation is also true. We shall illustrate these theorems by means of examples and show that the upper bound obtained therein is also best possible. (cf., [76], [77], [78]).

4.1 STURM-LIOUVILLE DIFFERENCE EQUATIONS WITH AN INDEFINITE WEIGHT-FUNCTION:

In the sequel we let b_n, $n = 0, 1, \ldots, m-1$ be any finite sequence of real numbers and $a_n \neq 0$, $n = 0, 1, \ldots, m-1$. Furthermore let $c_n > 0$, $n = -1, 0, 1, \ldots, m-1$ where $c_{-1} > 0$ is fixed. We shall be dealing with the formally self-adjoint second order difference equation

$$-\Delta(c_{n-1}\Delta y_{n-1}) + b_n y_n = \lambda a_n y_n \qquad n = 0, 1, \ldots, m-1 \quad (4.1.0)$$

where Δ is the forward difference operator, λ is a parameter and $m \geq 2$.

If we introduce the boundary conditions

$$y_{-1} = 0 \; , \; y_m = 0 \tag{4.1.1}$$

then (4.1.0-1) define an eigenvalue problem. A solution of (4.1.0-1) is then a (possibly complex) m-vector $y(\lambda)$,

$$y(\lambda) = \big(y_0(\lambda) \, , y_1(\lambda) \, , \ldots , y_{m-1}(\lambda)\big) \tag{4.1.2}$$

where $y_{-1}(\lambda) = y_m(\lambda) = 0$.

For a given m-vector $f = (f_0 \, , f_1 \, , \ldots , f_{m-1})$ we define the m-vector $\ell[f]$ by

$$\ell[f]_n = a_n^{-1}\{-\Delta(c_{n-1}\Delta f_{n-1}) + b_n f_n\} \tag{4.1.3}$$

where we take it that $f_{-1} = f_m = 0$ by definition. For given m-vectors f, g we define their "J-inner-product" by

$$[f \, , g] = \sum_0^{m-1} f_n \bar{g}_n a_n \; . \tag{4.1.4}$$

The following formula of summation by parts [30, p. 17] should also be useful,

$$\sum_M^N u_k \Delta v_k = \big[u_{k-1} v_k\big]_M^{N+1} - \sum_M^N v_k \Delta u_{k-1} \; . \tag{4.1.5}$$

We now define a quadratic functional $Q(f)$ with domain the collection of all complex m-vectors f by

$$Q(f) = [f, \ell[f]] \tag{4.1.6}$$

$$= c_{-1}|f_0|^2 + \sum_{0}^{m-1} \{c_n|\Delta f_n|^2 + b_n|f_n|^2\} \tag{4.1.7}$$

where we obtain (4.1.7) from (4.1.6) upon the application of (4.1.5).

<u>THEOREM 4.1.1</u>:

Let $y(\lambda) = (y_0(\lambda), y_1(\lambda), \ldots, y_{m-1}(\lambda))$ be a solution of (4.1.0) satisfying $y_{-1}(\lambda) = 0$. Then for $0 \leq n \leq m-1$,

$$(\bar{\mu} - \lambda) \sum_{r=0}^{n} a_r y_r(\lambda) y_r(\mu) = c_n \begin{vmatrix} y_{n+1}(\lambda) & \overline{y_{n+1}(\mu)} \\ y_n(\lambda) & \overline{y_n(\mu)} \end{vmatrix} .$$

<u>Proof</u>: This result can be proven as in [3, p. 98, Theorem 4.2.1] and so we omit the details.

<u>THEOREM 4.1.2</u>:

If λ is non-real, $0 \leq n \leq m-1$, then

$$\sum_{r=0}^{n} a_r |y_r(\lambda)|^2 = \frac{1}{2i \operatorname{Im} \lambda} c_n \begin{vmatrix} y_{n+1}(\lambda) & \overline{y_{n+1}(\lambda)} \\ y_n(\lambda) & \overline{y_n(\lambda)} \end{vmatrix} .$$

Proof: We refer to [3, p. 99, Theorem 4.2.3].

THEOREM 4.1.3:

Let $\tau_r = [y(\lambda_r), y(\lambda_r)]$. If λ_r , λ_s are non-real eigenvalues of (4.1.0-1) and $\bar{\lambda}_r \neq \lambda_s$ then

$$[y(\lambda_s), y(\lambda_r)] = 0 . \qquad (4.1.8a)$$

Hence

$$[y(\lambda_s), y(\lambda_r)] = \tau_r \delta_{rs} \qquad (4.1.8b)$$

where δ_{rs} is the Kronecker delta and when $r \neq s$ we mean that λ_r , λ_s are not conjugates.

Proof: Similar to [3, p. 104, Theorem 4.4.1].

THEOREM 4.1.4:

Let $\lambda_0, \lambda_1, \ldots, \lambda_{m-1}$; $f(\lambda_0), \ldots, f(\lambda_{m-1})$ denote the eigenvalues and corresponding eigenvectors of the problem

$$-\Delta(c_{n-1}\Delta y_{n-1}) + b_n y_n = \lambda y_n \qquad (4.1.9)$$

with the boundary conditions (4.1.1). Let y be an arbitrary m-vector. Then

$$y_n = \sum_{r=0}^{m-1} v(\lambda_r)\rho_r^{-1} f_n(\lambda_r) \qquad n = 0, 1, \ldots, m-1 \qquad (4.1.10)$$

where

$$v(\lambda_r) = \sum_{s=0}^{m-1} y_s f_s(\lambda_r) \qquad r = 0, 1, \ldots, m-1 \qquad (4.1.11)$$

and

$$\rho_r = \sum_{n=0}^{m-1} |f_n(\lambda_r)|^2 \ . \qquad (4.1.12)$$

Proof: This follows from the results in §4.4 of [3] where we make use of Theorem 4.4.2 (with $a_n = 1$).

THEOREM 4.1.5:

Let y be an arbitrary m-vector (with real or complex components) with $y_{-1} = y_m = 0$. Then

$$Q(y) = \sum_{n=0}^{m-1} \lambda_n |v(\lambda_n)|^2 \rho_n^{-1} \qquad (4.1.13)$$

where λ, v, ρ are as in Theorem 4.1.4.

Proof: Let us write $k_i \equiv v(\lambda_i)\rho_i^{-1}$ for brevity. Note also that

$$\ell[f(\lambda_s)]_n = \lambda_s a_n^{-1} f_n(\lambda_s) \ . \qquad (4.1.14)$$

Hence

$$Q(y) = \sum_{n=0}^{m-1} y_n \overline{\ell[y]_n}\, a_n$$

$$= \sum_{n=0}^{m-1} \left\{\sum_{r=0}^{m-1} k_r f_n(\lambda_r)\right\}\left\{\sum_{s=0}^{m-1} \lambda_s a_n^{-1} \bar{k}_s f_n(\lambda_s)\right\} a_n$$

where we have used the expansion (4.1.10) along with (4.1.14),

$$= \sum_{s=0}^{m-1} \lambda_s \bar{k}_s \sum_{r=0}^{m-1} v(\lambda_r) \sum_{n=0}^{m-1} \rho_r^{-1} f_n(\lambda_r) f_n(\lambda_s)$$

$$= \sum_{s=0}^{m-1} \lambda_s \bar{k}_s \sum_{r=0}^{m-1} v(\lambda_r)\, \delta_{rs}$$

by Theorem 4.1.3 with $a_n = 1$ (see also [3, p. 105, (4.4.2)]),

$$= \sum_{s=0}^{m-1} \lambda_s \bar{k}_s v(\lambda_s)$$

$$= \sum_{s=0}^{m-1} \lambda_s |v(\lambda_s)|^2 \rho_s^{-1}$$

which is what we set out to prove.

<u>COROLLARY 4.1.1</u>:

Let y be any m-vector. Then

$$c_{-1}|y_0|^2 + \sum_0^{m-1} \left\{c_n|\Delta y_n|^2 + b_n|y_n|^2\right\} = \sum_0^{m-1} \lambda_n |v(\lambda_n)|^2 \rho_n^{-1} \tag{4.1.15}$$

Proof: This follows from (4.1.7) and the preceding theorem.

LEMMA 4.1.1:

Let $N > 0$ be the number of distinct negative eigenvalues of the problem (4.1.9), (4.1.1) and denote these by $\lambda_0, \lambda_1, \ldots, \lambda_{N-1}$ with $f(\lambda_0), \ldots, f(\lambda_{N-1})$ representing the corresponding N eigenvectors.

Let there be an m-vector y which is orthogonal to the above collection of eigenvectors, i.e.

$$\sum_{r=0}^{m-1} y_r f_r(\lambda_s) = 0 \qquad s = 0, 1, \ldots, N-1 . \tag{4.1.16}$$

Then

$$Q(y) > 0 \tag{4.1.17}$$

if (4.1.9), (4.1.1) has at least one positive eigenvalue.

Proof: (4.1.16) implies that $v(\lambda_s) = 0$, $s = 0, \ldots, N-1$.

Hence

$$\begin{aligned} Q(y) &= \left\{ \sum_{n=0}^{N-1} + \sum_{n=N}^{m-1} \right\} \lambda_n |v(\lambda_n)|^2 \rho_n^{-1} \\ &= \sum_{n=N}^{m-1} \lambda_n |v(\lambda_n)|^2 \rho_n^{-1} \\ &> 0 \end{aligned}$$

since at least one of the $\lambda_N, \ldots, \lambda_{m-1}$ is positive. This completes the proof.

We note that 0 is an eigenvalue of (4.1.9), (4.1.1) if and only if it is an eigenvalue of (4.1.0-1) having the same multiplicity in both cases. On the other hand if (4.1.9), (4.1.1) has no positive eigenvalues then the number of pairs of non-real eigenvalues of (4.1.0-1) is necessarily less than or equal to the number of negative eigenvalues of (4.1.9), (4.1.1) since the defining relations for the eigenvalues are, in both cases, polynomials of degree m. Thus we shall always assume that (4.1.9), (4.1.1) have at least one positive eigenvalue.

Let M = the number of distinct pairs of non-real eigenvalues of (4.1.0-1). (By a pair we mean an eigenvalue and its conjugate.)

Let N = the number of distinct negative eigenvalues of (4.1.9), (4.1.1).

<u>THEOREM 4.1.6</u>:

Let M, N be defined as in the preceding Remark. Then

$$M \leqq N . \tag{4.1.18}$$

The upper bound is best possible.

Proof: Let $\mu_0, \mu_1, \dots, \mu_{M-1}; \bar{\mu}_0, \bar{\mu}_1, \dots, \bar{\mu}_{M-1}$ be the non-real eigenvalues of (4.1.0-1) and

$$y(\mu_i) = \left(y_0(\mu_i), y_1(\mu_i), \dots, y_{m-1}(\mu_i)\right)$$

be the eigenvector corresponding to the eigenvalue μ_i, where $0 \leqq i \leqq M-1$ (we note that $\bar{y}(\mu_i)$ is the eigenvector corresponding to $\bar{\mu}_i$).

We define an m-vector $z = (z_0, z_1, \dots, z_{m-1})$ where

$$z_i = \sum_{n=0}^{M-1} e_n y_i(\mu_n) \qquad i = 0, \dots, m-1 \tag{4.1.19}$$

and the e_n are to be chosen later. Because of the indexing of the eigenvalues we have

$$\bar{\mu}_i \neq \mu_j \qquad 0 \leqq i, j \leqq M-1 .$$

Consequently Theorem 4.1.3 implies that

$$[y(\mu_i), y(\mu_j)] = 0, \quad i \neq j \tag{4.1.20}$$

and

$$[y(\mu_i), y(\mu_i)] = 0$$

since the eigenvalues are non-real.

Suppose, if possible, that $M > N$. We shall proceed

to show that the e_n in (4.1.19) can be chosen so that z is orthogonal to $f_0, f_1, \ldots, f_{N-1}$.

Thus we must have

$$\sum_{n=0}^{m-1} z_n f_n(\lambda_r) = 0 \qquad r = 0, \ldots, N-1 .$$

Substituting (4.1.19) into the latter equation we find that the e_n must satisfy the following system of N equations in M unknowns,

$$\sum_{j=0}^{M-1} e_j \left\{ \sum_{n=0}^{m-1} f_n(\lambda_r) y_n(\mu_j) \right\} = 0$$

where $r = 0, 1, \ldots, N-1$. Since $M > N$ the last equation always has a non-trivial solution $e_0, e_1, \ldots, e_{M-1}$. Fix such a set of (e_i). The resulting m-vector z then has the property that

$$Q(z) > 0 \tag{4.1.21}$$

by Lemma 4.1.1. Moreover $z_{-1} = z_m = 0$. Hence (4.1.7) implies that

$$Q(z) = [z, \ell[z]]$$

$$= \sum_{n=0}^{m-1} z_n \overline{\ell[z]_n} a_n$$

$$= \sum_{n=0}^{m-1} \left\{\sum_{r=0}^{M-1} e_r y_n(\mu_r)\right\}\left\{\sum_{s=0}^{M-1} \bar{e}_s \bar{\mu}_s \bar{y}_n(\mu_s)\right\} a_n$$

$$= \sum_{r,s=0}^{M-1} e_r \bar{e}_s \bar{\mu}_s \left\{\sum_{n=0}^{m-1} a_n y_n(\mu_r) \bar{y}_n(\mu_s)\right\}$$

$$= \sum_{r,s=0}^{M-1} e_r \bar{e}_s \bar{\mu}_s [y(\mu_r) , y(\mu_s)] .$$

Now since $\bar{\mu}_r \neq \mu_s$ for all r , s , $0 \leqq r , s \leqq M-1$, (4.1.20) implies

$$[y(\mu_r) , y(\mu_s)] = 0$$

for all r , s , $0 \leqq r , s \leqq M-1$.

Hence

$$Q(z) = 0 . \qquad (4.1.22)$$

This, however, is in contradiction with (4.1.21). Thus $M \leqq N$ and the theorem is completely proved.

EXAMPLE 4.1.1:

Let $c_n = 1$, $b_n = -2$ and $a_n = (-1)^n$. Consider the problem

$$-\Delta^2 y_{n-1} - 2y_n = \lambda(-1)^n y_n \qquad n = 0 , \ldots , m-1 \qquad (4.1.23a)$$

$$y_{-1} = 0 = y_m \qquad (4.1.23b)$$

where $m \geqq 2$. The corresponding "definite" problem

$$-\Delta^2 y_{n-1} - 2y_n = \lambda y_n \tag{4.1.24}$$

with the same boundary conditions must have only real eigenvalues and consequently the zeros of $y_m(\lambda)$ are all real.

If we put $y_0 = 1$ a straightforward computation shows that, if $m = 2k$ then

$$y_{2k}(\lambda) = y_{2k}(-\lambda)$$

and thus $y_{2k}(\lambda)$ has an equal number, k , of positive and negative zeros which are precisely the eigenvalues of the associated definite problem. In fact $y_{2k}(\lambda)$ consists of only even powers of λ while $y_{2k+1}(\lambda)$ has only odd powers of λ .

The following table illustrates this:

TABLE I

	1	λ	λ^2	λ^3	λ^4	λ^5	λ^6	λ^7	λ^8	λ^9	
$y_0(\lambda)$	1										...
$y_1(\lambda)$	0	-1									...
$y_2(\lambda)$	-1	0	1								...
$y_3(\lambda)$	0	2	0	-1							...
$y_4(\lambda)$	1	0	-3	0	1						...
$y_5(\lambda)$	0	-3	0	4	0	-1					...
$y_6(\lambda)$	-1	0	6	0	-5	0	1				...
$y_7(\lambda)$	0	4	0	-10	0	6	0	-1			...
$y_8(\lambda)$	1	0	-10	0	15	0	-7	0	1		...
$y_9(\lambda)$	0	-5	0	17	0	-21	0	8	0	-1	...
...	...	...	...	...	...	...	...	...	...	...	...

The table gives the coefficients of the corresponding power of λ and of the polynomial concerned. To obtain the coefficients of the polynomial solutions of the indefinite problem (4.1.22-23) (up to a possible sign change) we simply change the negative signs in the above table to positive signs. The resulting table then shows that if $z_{2k}(\lambda)$ is the polynomial corresponding to $y_{2k}(\lambda)$ then the former has positive coefficients and only even powers of λ ; consequently its zeros must be non-real and occur in conjugate pairs. Thus there are $M = k$ pairs of non-real eigenvalues, if $m = 2k$, and $N = k$ negative eigenvalues of (4.1.24-23).

§4.2 STURM-LIOUVILLE DIFFERENTIAL EQUATIONS WITH AN INDEFINITE WEIGHT-FUNCTION

The main difference between the handling of eigenvalue problems for differential and difference equations, in the regular case, is that the latter always has at most a finite number of eigenvalues while the former always has an infinite number of eigenvalues, under proper conditions. In one case the "eigenfunction" expansion is trivial while in the other it is more involved. Still, the number of non-real eigenvalues is, in both cases, finite and bounded by similar constants.

In this section we shall make the following assumptions: That $p(x) > 0$ on $[a, b]$, $q(x) \in L_{loc}(a, b)$ is such that the problem

$$-\{p(x)y'\}' + q(x)y = \lambda y \tag{4.2.1}$$

$$y(a) = y(b) = 0 \tag{4.2.2}$$

admits a denumerable number of eigenvalues having no finite point of accumulation. (Conditions which guarantee this are that both p^{-1}, $q \in L(a, b)$, see [3, p. 215, Theorem 8.4.6].) In this case the eigenfunctions form a complete orthonormal set in $L^2(a, b)$.

Associated with (4.2.1-2) is the "indefinite" boundary problem

$$-\{p(x)z'\}' + q(x)z = \lambda r(x)z \tag{4.2.3}$$

$$z(a) = z(b) = 0 \tag{4.2.4}$$

where $r(x)$ is a real-valued function defined on $[a, b]$ and such that $r(x)$ takes both signs on some subsets of positive measure. The "*Indefinite case*" is characterized by the fact that both q, r have a variable sign in $[a, b]$ (see [53, p. 288] and not being equal a.e.

LEMMA 4.2.1:

Let f be an eigenfunction corresponding to some eigenvalue λ of (4.2.3-4). Then

$$\int_a^b (p|f'|^2 + q|f|^2)dx = \lambda \int_a^b r|f|^2\, dx . \tag{4.2.5}$$

Proof: We multiply (4.2.3) by $\bar{f}$ and integrate both sides over the interval $[a, b]$ to find

$$\int_a^b (pf')'\, \bar{f}\, dx + \lambda \int_a^b r|f|^2\, dx = \int_a^b q|f|^2\, dx .$$

Integrating the first integral by parts and applying the boundary conditions (4.2.4) the result follows.

LEMMA 4.2.2:

Let $\lambda, \mu, \lambda \neq \bar{\mu}$ be two non-real eigenvalues with eigenfunctions f, g respectively. Then

$$\int_a^b r(x)f(x)\bar{g}(x)dx = 0 \qquad (4.2.6)$$

i.e., f, g are J-orthogonal in the Krein space $L^2(|r|)$ and

$$\int_a^b \{p(x)f'(x)\bar{g}'(x) + q(x)f(x)\bar{g}(x)\}dx = 0 . \qquad (4.2.7)$$

<u>Proof:</u> We have

$$-(pf')' + qf = \lambda rf \qquad (4.2.8)$$

$$-(p\bar{g}')' + q\bar{g} = \bar{\mu}r\bar{g} \qquad (4.2.9)$$

along with $f(a) = f(b) = g(a) = g(b) = 0$. Multiplying (4.2.8) by $\bar{g}$ and (4.2.7) by f and subtracting the results we obtain, upon integration over $[a, b]$,

$$(\lambda - \bar{\mu}) \int_a^b r(x)f(x)\bar{g}(x)dx = \int_a^b \{(p\bar{g}')' f - (pf')' \bar{g}\}dx$$

and integrating the latter integral by parts we find

$$\int_a^b \{(p\bar{g}')' f - (pf')' \bar{g}\}dx = [p(\bar{g}'f - \bar{g}f')]_a^b$$

$$= 0$$

because of the boundary conditions. This proves (4.2.6) since $\lambda \neq \bar{\mu}$.

Multiplying (4.2.8) by $\bar{g}$ and integrating over $[a, b]$ we obtain

$$\int_a^b \{-(pf')'\bar{g} + qf\bar{g}\}dx = \lambda \int_a^b rf\bar{g}\, dx .$$

Integrating the first term in the left by parts we see that

$$-\int_a^b (pf')'\bar{g} = -[pf'\bar{g}]_a^b + \int_a^b pf'\bar{g}'\, dx$$
$$= \int_a^b pf'\bar{g}'\, dx .$$

Thus

$$\int_a^b \{pf'\bar{g}' + qf\bar{g}\}dx = \lambda \int_a^b rf\bar{g}\, dx$$
$$= 0$$

by (4.2.6). This completes the proof.

Associated with (4.2.1) is the differential operator A defined in $L^2(a, b)$ with domain $\mathcal{D}(A)$ defined by

$$\mathcal{D}(A) = \{y \in L^2(a, b) : y, py' \in AC_{loc}(a, b) \quad \text{and} \quad Ay \in L^2(a, b)\}$$

where for $y \in \mathcal{D}(A)$,

$$Ay = -(py')' + qy .$$

If we let $\mathcal{D}(\tilde{A})$ be defined by

$$\mathcal{D}(\tilde{A}) = \{y \in \mathcal{D}(A) : y(a) = y(b) = 0\} \qquad (4.2.10)$$

and let

$$\tilde{A}y = Ay \qquad y \in \mathcal{D}(\tilde{A}) \qquad (4.2.11)$$

then $\tilde{A}$ is a restriction of A to $\mathcal{D}(\tilde{A})$ and $\tilde{A}$ is, in fact, a symmetric operator [34, §4.11, Theorem 1].

The following lemmas are part of the theory of the regular Sturm-Liouville equation and can be found in [34], thus we omit the proofs.

<u>LEMMA 4.2.3</u>:

a) The regular Sturm-Liouville operator $\tilde{A}$, defined above, is bounded below, i.e. there exists a constant $\gamma \in \mathbb{R}$ such that

$$(\tilde{A}f , f) \geqq \gamma(f , f) , \quad f \in \mathcal{D}(\tilde{A}) \qquad (4.2.12)$$

where $(,)$ is the usual inner product in $L^2(a , b)$.

b) The operator $\tilde{A}$ has at most a finite number of negative eigenvalues.

<u>Proof</u>: For part a) see [34, §5.17, Ex. 5.3° and §6.7, Corollary]. Part b) is proved in [34, §5.8, Theorem 2].

For $f \in \mathcal{D}(\tilde{A})$, the expression $(\tilde{A}f , f)$ defines a quadratic functional with values

$$(\tilde{A}f\ ,\ f)\ =\ \int_a^b \{p|f'|^2 + q|f|^2\}dx \qquad f \in \mathcal{D}(\tilde{A})\ .$$

This is immediate if we follow the argument leading to (4.2.5).

<u>LEMMA 4.2.4</u>:

We define $D(Q)$ by

$$D(Q)\ =\ \{y \in L^2(a\ ,\ b)\ :\ \sum_0^\infty |\lambda_j||(y\ ,\ \phi_j)|^2 < \infty\}$$

$$Q(y)\ =\ \sum_0^\infty \lambda_j|(y\ ,\ \phi_j)|^2 \qquad y \in D(Q)$$

where (λ_j) , (ϕ_j) are the eigenvalues and eigenfunctions of (4.2.1-2) respectively.

Henceforth (,) will denote the inner product in $L^2(a\ ,\ b)$. Then the quadratic functional $Q(y)$ in $D(Q)$ is an extension of the functional $(\tilde{A}y\ ,\ y)$ in $\mathcal{D}(\tilde{A})$, i.e.

$$Q(y)\ =\ (\tilde{A}y\ ,\ y) \qquad y \in \mathcal{D}(\tilde{A})$$

or

$$\int_a^b \{p|y'|^2 + q|y|^2\}dx\ =\ \sum_0^\infty \lambda_j|(y\ ,\ \phi_j)|^2 \qquad (4.2.13)$$

for $f \in \mathcal{D}(\tilde{A})$.

<u>Proof</u>: The proof of this theorem is contained in [34, §6,

Theorem 1, pp. 6.1-6.5].

We now define another quadratic functional $Q'(y)$ with domain $D(Q')$ given by

$$D(Q') = \{y \in AC(a, b) : p|y'|^2 \in L(a, b) \text{ and } y(a) = y(b) = 0\} \tag{4.2.14}$$

and for $y \in D(Q')$

$$Q'(y) = \int_a^b \{p|f'|^2 + q|f|^2\} .$$

Then $Q'(y)$ is defined and extends $(\tilde{A}y, y)$ [34, p. 6.6]. The crux of the matter is the following lemma,

LEMMA 4.2.5:

When $p(x) > 0$ a.e. the extensions $Q(y)$, $Q'(y)$ of the quadratic functional $(\tilde{A}y, y)$ are identical, i.e. $D(Q) = D(Q')$ and

$$\int_a^b \{p|y'|^2 + q|y|^2\}dx = \sum_0^\infty \lambda_j |(y, \phi_j)|^2$$

where $y \in D(Q')$.

Proof: This important result is proved in [34, p. 6.8, Theorem 3

THEOREM 4.2.1: (cf., [76])

Let $p(x) > 0$ a.e. on $[a, b]$, $q \in L(a, b)$ and $r(x)$ as in the hypotheses following (4.2.3-4). The eigenvalue problem (4.2.3-4) possesses at most a finite number of non-real eigenvalues. If we let

M = the number of pairs of distinct non-real eigenvalues of (4.2.3-4),

N = the number of distinct negative eigenvalues of (4.2.1-2) (which we know is finite by Lemma 4.2.3(b)),

then

$$M \leqq N . \qquad (4.2.15)$$

Proof: We let $\lambda_0, \lambda_1, \ldots, \lambda_{N-1}$ be the negative eigenvalues of (4.2.1-2) arranged in an increasing order of magnitude. Let $\phi_0, \ldots, \phi_{N-1}$ be the corresponding eigenfunctions. If for some $f \in D(Q')$ we have $(f, \phi_j) = 0$, $j = 0, \ldots, N-1$, then

$$Q'(f) > 0 \qquad (4.2.16)$$

since

$$Q'(f) = \sum_{N}^{\infty} \lambda_j |(f, \phi_j)|^2$$

and the $\lambda_j > 0$.

We now let $\mu_0, \mu_1, \ldots, \mu_{M-1}$ ($\mathrm{Im}\,\mu_i \neq 0$) be the non-real eigenvalues of (4.2.3-4) arranged such that

$$\bar{\mu}_i \neq \mu_j \qquad 0 \leqq i , j \leqq M-1 .$$

We write the corresponding eigenfunctions as

$$z_0(x) , z_1(x) , \ldots , z_{M-1}(x) . \tag{4.2.17}$$

Then $z_i(x) \in D(Q')$ $i = 0, \ldots, M-1$ because of (4.2.7), upon replacing g by f . Thus (4.2.17) constitutes a collection of M mutually J-orthogonal (in the Kreȋn space $L^2(|r|)$) eigenfunctions.

Let $e_j \in \mathbb{C}$ and form the sum

$$f(x) = \sum_{0}^{M-1} e_j z_j(x) \tag{4.2.18}$$

where the e_j's shall be chosen later. Assume, if possible that $M > N$.

Then it is possible to choose the (e_j) so that f is orthogonal (in the $L^2(a, b)$-sense) to $\phi_0, \ldots, \phi_{N-1}$. For it is necessary that

$$(f, \phi_j) = 0 \qquad j = 0, 1, \ldots, N-1$$

and so

$$\sum_{i=0}^{M-1} e_i(z_i, \phi_j) = 0, \quad j = 0, 1, \ldots, N-1 .$$

The latter constitutes a set of N linear equations in M unknowns where $M > N$. Thus this system has a non-trivial solution (e_j) not all zero which we fix. It is then necessary that, for such a choice of (e_j),

$$Q'(f) > 0 \tag{4.2.19}$$

because of a preceding remark. Moreover,

$$\begin{aligned} Q'(f) &= Q'\{\Sigma e_j z_j\} \\ &= \int_a^b p(\Sigma e_j z_j')(\Sigma \bar{e}_i \bar{z}_i') + q(\Sigma e_j z_j)(\Sigma \bar{e}_i \bar{z}_i) \\ &= \sum_{i,j=0}^{M-1} e_j \bar{e}_i \int_a^b \{p z_j' \bar{z}_i' + q z_j \bar{z}_i\} dx . \end{aligned}$$

But since $\mu_i \neq \bar{\mu}_j$ for all $0 \leq i, j \leq M-1$, Lemma 4.2.2 applies and so

$$\int_a^b \{p z_j' \bar{z}_i' + q z_j \bar{z}_i\} dx = 0$$

for all i, j, $0 \leq i, j \leq M-1$. Consequently

$$Q'(f) = 0$$

which contradicts (4.2.19). Thus $M \leq N$ and the theorem is completely proved.

The preceding theorem along with the results of Richardson [53] mentioned in the introduction to this chapter make the following notions plausible. Consider the equation

$$Ly \equiv -(py')' + qy = \lambda ry \tag{4.2.20}$$

where $p(x) > 0$ and continuous on $[a, b]$, $q(x)$ is continuous and is negative in a subinterval of $[a, b]$ and $r(x)$ is continuous on $[a, b]$ with the property that $r(x)$ changes sign at least once in $[a, b]$. In this case the space $L^2(|r|; [a, b]) = H$ defined by those (equivalence classes of) functions f such that

$$(f, f) \equiv \int_a^b |f|^2 |r| dx < \infty$$

is a Krein space with the indefinite inner product given by

$$[f, f] \equiv \int_a^b |f|^2 r\, dx \qquad f \in H$$

(see Appendix III and the references therein for more discussion on these spaces).

Let $U_1(y)$, $U_2(y)$ be the linear forms defined in Appendix I.4, equation (I.4.1). We denote the relationships

$$U_1(y) = 0$$

$$U_2(y) = 0$$

by

$$U(y) = 0 \ . \qquad (4.2.21)$$

The problem

$$\pi : \ Ly = \lambda ry \quad , \quad Uy = 0 \qquad (4.2.22)$$

then defines an eigenvalue problem, i.e., we seek values of $\lambda \in \mathbb{C}$ such that (4.2.22) has a non-trivial solution satisfying (4.2.21).

With some loss of generality, we shall say that the eigenvalue problem π is *formally* J-*self-adjoint* if

$$[f , Lg] = [Lf , g]$$

for all $f , g \in C^2(a , b)$ which satisfy

$$U(f) = U(g) = 0 \ .$$

For a J-self-adjoint problem, non-real eigenvalues may or may not exist but, in any case, if they do exist, their number appears to be finite for general boundary conditions also, because of the preceding theorem. Combining the results in [53, §4] with the preceding theorem we can formulate

<u>THEOREM 4.2.2</u>:

The *formally J-self-adjoint* problem

$$Ly = \lambda ry \qquad y(a) = y(b) = 0$$

has *a finite* number of non-real eigenvalues, *in some cases none at all*, and on $|\lambda| > \Lambda$ has only real eigenvalues, with no finite point of accumulation, clustering at minus infinity and plus infinity.

The second part of this theorem is due to Richardson [53, p. 301, Theorem VII]. It would seem plausible that Theorem 4.2.2 remains true for arbitrary "J-self-adjoint" boundary conditions though we shall not go into this at the present time. Theorem 4.2.1 extends, with appropriate changes in the argument, to the general even order formally self-adjoint differential equation

$$(-1)^n \left(p_0 y^{(n)}\right)^{(n)} + (-1)^{n-1} \left(p_1 y^{(n-1)}\right)^{(n-1)} + \cdots + p_n y = \lambda ry$$

$$y^{(j)}(a) = y^{(j)}(b) = 0 \qquad j = 0 , \ldots , n-1$$

where $p_0 > 0$ and $p_i(x)$ changes sign at least once in $[a , b]$, where i is in the range $0 < i \leqq n$ and we assume that $p_k \in C^{(n-k)}(a , b)$ (see Appendix III.4).

CHAPTER 5

INTRODUCTION:

In this chapter we study the spectrum of generalized differential operators generated by the expression

$$\ell[y](x) = -\frac{d}{d\nu(x)}\left\{y'(x) - \int_a^x y(s)d\sigma(s)\right\} \qquad x \in I \quad (5.0.0)$$

where $I = [a, \infty)$, ν is non-decreasing, σ of bounded-variation locally, ν , σ satisfying the basic assumptions of Chapter 3. We shall begin by proving a lemma which is similar in content to a theorem of Glazman [23] and the purpose of which is to relate the non-oscillatory behaviour of solutions of

$$\ell[y](x) = \lambda y(x) \qquad x \in I \qquad (5.0.1)$$

to the finiteness of the spectrum to the left of λ . We can then use some non-oscillation results from Chapter 2 to obtain criteria for the discreteness of the negative part of the spectrum of any self-adjoint extension of the minimal operator, which we shall define later. We shall apply the latter result and then obtain an extension of a theorem of M. Sh. Birman

[23, p. 93] gives a necessary and sufficient condition for the discreteness of the spectrum of second-order differential operators in the polar case. We shall then study the continuous spectrum (also called the essential spectrum) of these generalized operators particularly in the case when σ is of bounded variation on all of I. In the latter case we shall show that the corresponding self-adjoint extension is bounded below and give an explicit lower bound which extends a result of Everitt [16] in the case $p = 1$.

We note that since ν is non-decreasing the space $L^2(\nu, I)$ is a Hilbert space, consequently the general theory of operators in these spaces shall be used though we shall not always give the details: These can either be found in [23] or the proofs, in the more general setting, are adaptable and so will be omitted. Moreover we shall assume that the resulting operators are single-valued. Conditions for this can be found in [35], [36]. In the applications the operators thus defined are indeed single-valued and so there is no great loss of generality in assuming this property in general.

§5.1 THE DISCRETE SPECTRUM OF GENERALIZED DIFFERENTIAL OPERATORS:

We shall be dealing with the expression

$$\ell[y](x) = -\frac{d}{d\nu(x)}\left\{y_+'(x) - \int_a^x y(s)\,d\sigma(s)\right\} \qquad x \in I \qquad (5.1.0)$$

where $I = [a, \infty)$, ν is a right-continuous non-decreasing function and σ is right-continuous and of bounded variation locally on I . In case the interval I is finite we assume, as usual, that both ν , σ are continuous at the end-points.

In the following, $AC[a , b]$ $(AC_0[a , b])$ will stand for the space of absolutely continuous functions (having compact support in $[a , b]$) on $[a , b]$. By a finite function we shall mean a function which vanishes identically outside some finite interval.

We define the spaces $S(\alpha , \beta)$ and $T(\alpha , \beta)$ by

$$S(\alpha , \beta) = \{z(x) \in AC[\alpha , \beta] : z' \in L^2(\alpha , \beta) \text{ and } z(\alpha) = z(\beta) = 0\}$$

$$\begin{aligned} T(\alpha , \beta) = \{y(x) \in AC[\alpha , \beta] : {}& y_+'(x) \text{ exists everywhere on } \\ & [\alpha , \beta] , \ y_+'(x) \text{ is } \nu\text{-absolutely continuous on} \\ & [\alpha , \beta] , \ dy_+'/d\nu \in L^2(\nu ; [\alpha , \beta]) \text{ and} \\ & y(\alpha) = y(\beta) = y'(\alpha) = y'(\beta) = 0\} . \end{aligned}$$

Note that

$$T(\alpha , \beta) \subset S(\alpha , \beta) .$$

The first result is similar to an important result of Krein (see the book of Glazman [23, p. 35, Lemma 5]).

<u>LEMMA 5.1.0</u>:

Let $z(x) \in S(\alpha , \beta)$ be a finite function with support $[\alpha , \beta]$. Then for any $\varepsilon > 0$, there exists a finite function

$y(x) \in T(\alpha, \beta)$ having the same support as $z(x)$ and such that

$$|Q[z] - Q[y]| < \varepsilon \tag{5.1.1}$$

where

$$Q[z] \equiv \int_a^\infty \{|z'|^2 dx + |z|^2 d\sigma(x)\} . \tag{5.1.2}$$

<u>Proof:</u> We shall adapt the ideas in [23] to this situation. This said, it will be sufficient to prove the lemma in the case where $\sigma \equiv$ constant, or for the quadratic functional $Q'[z]$ defined by

$$Q'[z] \equiv \int_a^\infty |z'|^2 dx . \tag{5.1.3}$$

This is because the fundamental nature of any sequence in the "Q-metric" can be realized by its nature in the "Q'-metric", i.e. If $\{y_j(x)\}$ is any sequence in $AC[\alpha, \beta]$ with $y_j(\alpha) = 0$ for all j,

$$\begin{aligned} |y_j(x) - y_k(x)| &\leqq \int_\alpha^x |y_j' - y_k'| dx \\ &\leqq \int_\alpha^\beta |y_j' - y_k'| dx \\ &\leqq (\beta - \alpha)^{\frac{1}{2}} \left\{ \int_\alpha^\beta |y_j' - y_k'|^2 dx \right\}^{\frac{1}{2}} . \end{aligned} \tag{5.1.4}$$

Thus e.g. If $y_j' \to y$ in the Q'-metric then y_j converges uniformly to some function on $[\alpha, \beta]$. We can define an inner product on the space F of all functions $z \in S(\alpha, \beta)$ which are finite with support $[\alpha, \beta]$ by

$$(f, g) \equiv \int_a^\infty f'(x)\bar{g}'(x)dx \qquad f, g \in F . \qquad (5.1.5)$$

This is well-defined and $Q'[f] = (f, f)$. The inner product then induces a norm $\| \|_F$ defined by $\|f\|_F = (f, f)^{\frac{1}{2}}$ such that the resulting metric space is complete. To prove this let $\{z_j(x)\}$ be a Cauchy sequence in F. Then if we write $z_j' \equiv w_j$ we find that

$$\|w_j - w_k\|_2 = \|z_j - z_k\|_F < \varepsilon$$

for j, k large, where $\| \|_2$ is the L^2-norm on (α, β). The completeness of $L^2(\alpha, \beta)$ then implies the existence of a function $h(x)$ such that

$$\lim_{j\to\infty} z_j'(x) = h(x)$$

in the norm of $L^2(\alpha, \beta)$.

We now set

$$z(x) = \int_\alpha^x h(s)ds .$$

Then $z \in AC[\alpha, \beta]$, $z(\alpha) = 0$ and $z' \in L^2(\alpha, \beta)$. Using

(5.1.4), with y replaced by z, we see that the sequence z_j converges uniformly on $[\alpha, \beta]$ so that

$$\lim_{j\to\infty} z_j(x) = z(x) \qquad x \in [\alpha, \beta]$$

and thus

$$z(\beta) = \lim_{j\to\infty} z_j(\beta) = 0 .$$

Hence $z \in F$. Thus F is complete relative to the norm $\| \|_F$, and so F is a Hilbert space. We now proceed to show that the collection G of functions in $T(\alpha, \beta)$ which are finite and have support $[\alpha, \beta]$ is dense in F. To prove this it suffices to show that if there is some function $z \in F$ which is F-orthogonal to all of G, then $z \equiv 0$. Let z be such a function. Then

$$\int_\alpha^\beta z'(x)\bar{y}'(x)dx = 0 \quad \text{for all} \quad y \in G .$$

Now an integration by parts shows that

$$\begin{aligned}\int_\alpha^\beta z'(x)\bar{y}'(x)dx &= \int_\alpha^\beta \bar{y}'(x)dz(x)\\ &= [\bar{y}'z]_\alpha^\beta - \int_\alpha^\beta z(x)d\bar{y}'(x)\\ &= -\int_\alpha^\beta z(x)d\bar{y}'(x)\\ &= -\int_\alpha^\beta z(x)\frac{d\bar{y}'(x)}{d\nu(x)}d\nu(x)\end{aligned}$$

because y' is ν-absolutely continuous. Hence we have that

$$\int_\alpha^\beta z(x)\left\{-\frac{d\bar{y}'(x)}{d\nu(x)}\right\}d\nu(x) = 0 \qquad (5.1.6)$$

for all $y \in G$. We now note that the operator L_0 with domain G defined by

$$L_0 y = -\frac{dy'}{d\nu} \qquad y \in G,$$

is in fact the minimal operator associated with the differential expression

$$\ell[y](x) = -\frac{dy'}{d\nu}(x) \qquad x \in [\alpha, \beta]$$

(see Chapter 3.6) and so is a closed symmetric operator with deficiency indices (2, 2). Thus (5.1.6) implies that

$$(L_0 y, z)_\nu = (y, 0)_\nu \qquad (5.1.7)$$

for all $y \in G$. Thus z is in the domain of the adjoint L_0^* which is equal to the "maximal" operator L defined in $L^2(\nu; [\alpha, \beta])$ (see Chapter 3.6, and [46, §17.3, Theorem 1]). Hence z is absolutely continuous on $[\alpha, \beta]$, $z_+'(x)$ exists everywhere on $[\alpha, \beta]$ and is ν-absolutely continuous there. Moreover from (5.1.7) we see that

$$L_0^* z = Lz = -\frac{dz'(x)}{d\nu(x)} = 0 \qquad (5.1.8)$$

for all $x \in [\alpha, \beta]$. Integrating with respect to ν we see that (5.1.8) implies that

$$z'(x) = \text{constant}$$

and so $z(x)$ must be linear on $[\alpha, \beta]$. But since $z \in F$, $z(\alpha) = z(\beta) = 0$. Hence

$$z(x) \equiv 0$$

and so G is dense in F . Thus for given $z \in F$, $\varepsilon > 0$, there exists $y \in G$ such that

$$\|z - y\|_F < \varepsilon$$

i.e.

$$|Q'[z] - Q'[y]| < \varepsilon \ .$$

Let $\varepsilon > 0$ be given, thus for given $z \in F$ there exists a $y \in G$ such that

$$|Q'[z] - Q'[y]| < \frac{\varepsilon}{2} \ .$$

Hence,

$$|Q[z] - Q[y]| \leqq |Q'[z] - Q'[y]| + \left|\int_\alpha^\beta (|z|^2 - |y|^2)d\sigma\right|$$

$$< \frac{\varepsilon}{2} + \int_\alpha^\beta \left||z|^2 - |y|^2\right| |d\sigma(x)| \ .$$

Since

$$|z(x) - y(x)| \leqq c\|z - y\|_F ,$$

by (5.1.4), $|y(x)| = O(|z(x)|)$ uniformly for $x \in [\alpha, \beta]$. Thus

$$\int_\alpha^\beta \left| |z|^2 - |y|^2 \right| |d\sigma| \leqq \left\{\int_\alpha^\beta \left| |z| - |y| \right|^2 |d\sigma|\right\}^{\frac{1}{2}} \left\{\int_\alpha^\beta \left| |z| + |y| \right|^2 |d\sigma|\right\}^{\frac{1}{2}}$$

$$\leqq c'\left\{\int_\alpha^\beta |z - y|^2 |d\sigma|\right\}^{\frac{1}{2}}$$

$$\leqq cc' \cdot \|z - y\|_F \cdot \left\{\int_\alpha^\beta |d\sigma|\right\}^{\frac{1}{2}} .$$

Hence we may restrict y further, if necessary, to make the right hand side of the inequality small. The result now follows and this completes the proof.

<u>COROLLARY 5.1.0</u>:

Let J consist of the collection of all finite functions f with support $[\alpha, \beta]$ such that $f \in AC[\alpha, \beta]$, $f' \in BV[\alpha, \beta]$ and $f(\alpha) = f(\beta) = 0$. Then

$$G \subset J \subset F \qquad (5.1.9)$$

and so J is dense in F.

<u>Proof</u>: If $f \in G$, then f' is ν-absolutely continuous and

so of bounded variation on $[\alpha, \beta]$. Thus $f \in \mathcal{J}$. On the other hand, for $f \in \mathcal{J}$, $f' \in BV[\alpha, \beta]$ and so $f' \in L^2(\alpha, \beta)$. The result now follows since $\mathcal{G}$ is dense in $\mathcal{F}$ by the preceding theorem.

COROLLARY 5.1.1:

Let $z \in \mathcal{F}$. Then there exists a function $y \in \mathcal{J}$ such that, for $\varepsilon > 0$,

$$|Q[z] - Q[y]| < \varepsilon . \tag{5.1.10}$$

Proof: This follows immediately from the preceding discussion.

LEMMA 5.1.1:

Let $\ell[y](x) = \lambda y(x)$ be non-oscillatory for $\lambda = \lambda_0$. Then for $z \in \mathcal{G}$,

$$Q[z] \geqq \lambda_0 \int_a^\infty |z|^2 d\nu . \tag{5.1.11}$$

Proof: We can assume that $\lambda_0 = 0$. Since $\ell[y](x) = 0$ is non-oscillatory, then

$$y'(x) = c + \int_a^x y(s)d\sigma(s) \qquad x \in [a, \infty) . \tag{5.1.12}$$

and so the latter equation is non-oscillatory. Thus there exists a solution $y(x)$ of (5.1.12) such that $y(x) \neq 0$ for all $x \geqq x_0$. Theorem 1.1.0 then implies that, for any $X \geqq x_0$, there is no function u such that $u \in AC[x_0, X]$, $u' \in BV(x_0, X)$, $u(x_0) = u(X) = 0$, and $Q[u] \leqq 0$. (For if there were one such then $y(x)$ would have a zero in (x_0, X) .) Thus for every such u , $Q[u] \geqq 0$. Since such u can be regarded as elements of J , with α, β replaced by x_0 , X respectively, then $Q[u] \geqq 0$ for all $u \in J$ and thus for all $u \in G$ by (5.1.9).

COROLLARY 5.1.2:

Suppose that (5.1.11) holds for all $z \in G$ and for $\lambda_0 = 0$. Then the spectrum of any self-adjoint extension of the minimal operator L_0 is finite to the left of $\lambda = 0$.

Proof: This follows from [23, pp. 34-35, Theorem 28].

THEOREM 5.1.0:

A necessary and sufficient condition for

$$\ell[y](x) = \lambda y(x) \tag{5.1.13}$$

to be oscillatory for $\lambda = \lambda_0$ is that the part of the spectrum of any self-adjoint extension of the minimal operator, lying to the left of λ_0 be an infinite set.

Proof: The argument is similar to that in [28, p. 40,

Theorem 31]. We can assume that $\lambda_0 = 0$. If (5.1.13) is oscillatory for $\lambda_0 = 0$ then for $x = t_1$ there exists a solution $y_1(x)$ of (5.1.12) which vanishes at α , β where $\beta > \alpha > t_1$. Then on $[\alpha , \beta]$ $y_1(x)$ can be regarded as an eigenfunction corresponding to $\lambda = 0$ of the problem

$$\ell[y] = 0$$

$$y(\alpha) = y(\beta) = 0 \; . \tag{5.1.14}$$

If we let $\alpha_1 = \alpha$ and $\beta_1 > \beta$, then from variational principles it follows that (5.1.14) along with

$$y(\alpha_1) = y(\beta_1) = 0$$

should have a negative eigenvalue λ_1 . Writing $f_1(x)$ for the corresponding eigenfunction we see that

$$Q[f_1] = \lambda_1(f_1 , f_1)_\nu < 0$$

where $(\, , \,)_\nu$ is the inner product in the space $L^2(\nu)$. Thus applying Lemma 5.1.0 we can find a function ϕ_1 in G such that

$$(L_0\phi_1 , \phi_1)_\nu = Q[\phi_1] < 0 \; .$$

Choosing $t_2 > \beta_1$ we can iterate the construction and we eventually obtain an infinite sequence of finite functions

$\phi_k \in \mathcal{D}(L_0)$ with disjoint supports such that

$$(L_0\phi_k , \phi_k)_\nu < 0 , \qquad k = 1 , 2 , \ldots .$$

Applying now Theorem 13 of [23, p. 15] we find that the negative part of the spectrum is an infinite set.

Conversely let us suppose that (5.1.13) is non-oscillatory for $\lambda_0 = 0$. Then Lemma 5.1.1 along with Corollary 5.1.2 imply that the spectrum to the left of 0 must be finite. This completes the proof.

As an application we obtain Theorem 31 of [23, §2.12] and Theorem 32 of [23, §2.14]. (For this reduction see the methods of Chapter 3.) It also follows, from the theorem just proved, that if (5.1.13) is oscillatory for $\lambda = \lambda_0$ then it is oscillatory for all $\lambda > \lambda_0$. Hence one of three cases must occur:

1) It is non-oscillatory for all λ .

2) It is oscillatory for all λ .

3) There is some λ_0 such that for $\lambda > \lambda_0$ it is oscillatory and for $\lambda < \lambda_0$ it is non-oscillatory.

This also applies to three-term recurrence relations.

<u>THEOREM 5.1.2</u>:

Let σ satisfy the usual hypotheses and suppose that

$\sigma(t)$ tends to a finite limit at infinity which we can assume is zero. Suppose furthermore that

$$\lim_{t\to\infty} t|\sigma(t)| = 0 . \qquad (5.1.15)$$

Let ν be non-decreasing. Then a necessary and sufficient condition for the spectrum of (5.1.13) to be discrete is that

$$\lim_{t\to\infty} t\bigl(\nu(\infty) - \nu(t)\bigr) = 0 \qquad (5.1.16)$$

when $\nu(\infty) < \infty$.

<u>Proof</u>: The spectrum will be discrete if and only if (5.1.13) is non-oscillatory for all λ . Moreover the latter holds if and only if the integro-differential equation

$$y'(t) = c + \int_a^t y(s)d\bigl(\sigma(s) - \lambda\nu(s)\bigr) \qquad (5.1.17)$$

is non-oscillatory for all λ . We can assume that $\nu(\infty) = 0$, since $\nu(\infty) < \infty$. Thus we need only show that under (5.1.15),

$$\lim_{t\to\infty} t|\nu(t)| = 0 \qquad (5.1.18)$$

if and only if (5.1.17) is non-oscillatory for all λ .
For given λ , choose t so large that

$$t|\sigma(t)| + |\lambda|t|\nu(t)| \leqq \frac{1}{4} \qquad t \geqq T .$$

Then

$$t|\sigma(t) - \lambda\nu(t)| \leqq \frac{1}{4} \qquad t \geqq T$$

and consequently Theorem 2.1.4 implies that (5.1.17) is non-oscillatory for such λ . On the other hand let (5.1.15) hold and suppose that (5.1.17) is non-oscillatory for all λ . Suppose that, on the contrary,

$$\lim_{t\to\infty} t|\nu(t)| \equiv \alpha \neq 0 .$$

By our hypothesis $\nu(t) < 0$ and so

$$t\left(\sigma(t) - \lambda\nu(t)\right) = t\sigma(t) + \lambda t|\nu(t)| .$$

We now choose t so large that

$$t\sigma(t) > -\frac{\alpha}{2} \qquad t \geqq T$$

and

$$t|\nu(t)| > \frac{\alpha}{2} \qquad t \geqq T .$$

Then, for $t \geqq T$, $\lambda > 0$,

$$t\left(\sigma(t) - \lambda\nu(t)\right) > \frac{\alpha}{2}(\lambda - 1) .$$

Since by hypothesis (5.1.17) is non-oscillatory for all λ we can choose λ so large that

$$\frac{\alpha}{2}(\lambda - 1) > \frac{1}{4} + \varepsilon$$

where $\varepsilon > 0$ is some fixed number. Thus for such a choice of λ

$$t\{\sigma(t) - \lambda\nu(t)\} > \frac{1}{4} + \varepsilon \qquad t \geqq T .$$

Thus $\sigma(t) - \lambda\nu(t)$ if positive for $t \geqq T$, such λ. An application of Theorem 2.2.1 shows that (5.1.17) is oscillatory for such λ. This is a contradiction and thus $\alpha = 0$. This completes the proof.

In particular when $\sigma(t) \equiv 0$, we obtain the result of Kac and Kreĭn [38, p. 78, Proposition 11.9°]. {For the original see p. 97, (2), of [38].} Again (5.1.15) is not superfluous. The latter result had extended a theorem of Birman [23, p. 93, Theorem 7] since we can let ν be absolutely continuous and then, when $\sigma(t) \equiv 0$, (5.1.13) is equivalent to

$$-y'' = \lambda\rho(x)y \qquad x \in [a, \infty) \tag{5.1.19}$$

where $\rho(x) > 0$. Glazman [23, §29] calls this case the "polar" case though the latter is usually connected with the sign indefiniteness of $\rho(x)$ in (5.1.19). Because of Theorem 5.1.0, other criteria for the finiteness of the negative part of the spectrum can be obtained via the non-oscillation theorems of Chapter 2.1. Moreover because of the applications to recurrence relations, we therefore obtain some criteria for the finiteness of the negative part of the spectrum of difference operators.

Example 1: If we let $\sigma(t) \equiv 0$ and define $\nu(t)$ by (3.8.3) and let $c_n = 1$ for all n or $t_n = n$ for all n, then (5.1.0) includes the difference equation

$$-\Delta^2 y_{n-1} = \lambda a_n y_n \qquad n = 0, 1, \ldots \tag{5.1.20}$$

where $a_n > 0$ by hypothesis. The discrete analog of Birman's theorem (above) is that the spectrum of (5.1.20) is discrete if and only if

$$\lim_{n\to\infty} n \sum_{j=n}^{\infty} a_j = 0$$

whenever $\Sigma a_n < \infty$. The proof follows the usual substitutions in (5.1.16) and is therefore omitted.

Example 2: In the addenda to Chapter 2 we saw that

$$-\Delta^2 y_{n-1} = \lambda y_n \qquad n = 0, 1, \ldots \tag{5.1.21}$$

is non-oscillatory whenever $\lambda \leqq 0$. Consequently, the above theory implies that the spectrum of (5.1.21) is finite below zero. Thus it shares the same properties, in this respect, as

$$-y'' = \lambda y \quad \text{on} \quad [a , \infty) \ .$$

Other criteria for the discreteness of differential and difference operators can be obtained from Theorem 2.3.4.

§5.2 THE CONTINUOUS SPECTRUM OF GENERALIZED DIFFERENTIAL OPERATORS:

In this section we study the continuous spectrum of the generalized differential equation

$$\ell[y](x) = -\frac{d}{dx}\left\{y'(x) - \int_a^x y\,d\sigma\right\} \qquad x \in I \tag{5.2.0}$$

where $I = [0 , \infty)$, and $\nu(x) = x$ so that equality is "almost everywhere" in the usual sense.

We shall hereafter assume that σ is of bounded variation over all of $[0 , \infty)$. Thus

$$\int_0^\infty |d\sigma(x)| < \infty \ . \tag{5.2.1}$$

If

$$\ell[y](x) = \lambda y(x) \qquad x \in I \tag{5.2.2}$$

then y satisfies the Volterra-Stieltjes integral equation

$$y(x) = y(0) + xy'(0) + \int_0^x (x-s)y(s)d\big(\sigma(s) - \lambda s\big) \qquad x \in I \tag{5.2.3}$$

(by the results in Chapter 3).

<u>THEOREM 5.2.1</u>: (Atkinson [3, p. 392], Theorem 12.6.1)

Let $\sigma(x)$ satisfy (5.2.1). Then (5.2.3) cannot have any positive eigenvalues. Hence (5.2.2) is limit-point at infinity.

<u>Proof</u>: For when $\lambda > 0$, (5.2.3) has no solution in $L^2(0, \infty)$ by virtue of Theorem 12.6.1 in [3]. Hence no eigenfunction can exist and thus the discrete spectrum is contained in $(-\infty, 0]$.

Defining the minimal operator L_0 corresponding to (5.2.0) as in Chapter 3, there is then a self-adjoint extension $\tilde{L}_0$ of L_0 which is determined by a set of homogeneous boundary conditions at 0 (see for example [46, §17.5]). Now if $\lambda < 0$, in (5.2.2), then Theorem 12.5.1 of [3, p. 384] implies that (5.2.3) has a pair of solutions y_1, y_2 admitting the asymptotic representations,

$$y_1(x) \sim \exp(-x\sqrt{|\lambda|})$$

$$y_2(x) \sim \exp(x\sqrt{|\lambda|}) .$$

It follows from this that y_1, y_2 are both eventually of constant sign and so, when $\lambda < 0$, (5.2.2) is non-oscillatory. Consequently the spectrum is finite on $(-\infty, 0]$ by Theorem 5.1.0. Thus there can be no point of the continuous spectrum in $(-\infty, 0)$. Hence

<u>THEOREM 5.2.2:</u>

Let σ satisfy (5.2.1). Then the continuous spectrum of (5.2.2) is precisely the semi-axis $[0, \infty)$.

From this it follows that the spectrum is bounded below and consequently any self-adjoint extension of the minimal operator is bounded below. Thus, for example, the self-adjoint operator generated by the differential expression

$$\ell[y] = \lambda y$$

and

$$y(0)\cos\alpha - y'(0)\sin\alpha = 0 \tag{5.2.4}$$

is bounded below, i.e. If we denote such an extension by L_α then

$$(L_\alpha f, f) \geqq -\gamma(f, f) \qquad f \in D_\alpha \tag{5.2.5}$$

where $\gamma \in \mathbb{R}$ and $(\, , \,)$ is the usual L^2 inner-product. We will now proceed to give an explicit lower bound γ for (5.2.5) in terms of σ . The approach used here is essentially an extension of an argument of Everitt [16]. The following lemma will be useful.

LEMMA 5.2.1: (Ganelius [22])

Let $f \geqq 0$ and g be functions of bounded variation on the closed interval J. Then

$$\int_J f\,dg \leqq \left\{\inf_J f + \operatorname{var}_J f\right\} \sup_{K\subset J} \int_K dg \tag{5.2.6}$$

where

$$\operatorname{var}_J f \equiv \int_J |df(x)|$$

and the sup is taken over all compact subsets of J. We recall the following notions:

The maximal domain $\mathcal{D}$ of the operator L generated by (5.2.0) in $L^2(0,\infty)$ is defined by

$$\mathcal{D} = \{f \in L^2(0,\infty) : f \in AC_{loc}(0,\infty),\ f'_+(x) \text{ exists on } [0,\infty),\ F(x) \in AC_{loc}(0,\infty), \text{ and } F'(x) \in L^2(0,\infty)\}$$

where

$$F(x) \equiv f'_+(x) - \int_a^x f(s)\,d\sigma(s) .$$

For $f \in D$,

$$Lf = \ell[f]$$

and L is a single-valued operator in $L^2(0, \infty)$. We denote the self-adjoint operator generated by (5.2.2) and $y(0) = 0$ by T_0. Thus

$$\mathcal{D}(T_0) = \{f \in \mathcal{D} : f(0) = 0\} \tag{5.2.7}$$

and

$$T_0 f = \ell[f] .$$

LEMMA 5.2.2:

Let σ satisfy (5.2.1). Then for every $\varepsilon > 0$ there exists a $C = C(\varepsilon) > 0$ such that

$$\int_0^x |f(t)|^2 \, |d\sigma(t)| \leqq C(\varepsilon) \cdot \int_0^x |f|^2 \, dt + \varepsilon \int_0^x |f'|^2 \, dt \tag{5.2.8}$$

for all $f \in \mathcal{D}$, and $x \in [1, \infty)$.

Proof: We use Lemma 5.2.1 with f, g replaced by $|f|^2$ and the variation of σ over $[0, x]$ respectively. Thus

$$\int_0^x |f(t)|^2 \, |d\sigma(t)| \leqq \left\{ \inf_{[0,x]} |f|^2 + \operatorname*{var}_{[0,x]} |f|^2 \right\} \int_0^x |d\sigma(t)| . \tag{5.2.9}$$

Now if $x \in [1, \infty)$ then

$$\inf_{[0,x]} |f|^2 \leqq \int_0^x |f|^2 \, dt . \tag{5.2.10}$$

Moreover,

$$\operatorname*{Var}_{[0,x]} |f|^2 = \int_0^x \left|d|f|^2\right| = \int_0^x 2|\mathrm{re}(ff')|dt$$

$$\leqq 2\left\{\int_0^x |f|^2 dt\right\}^{\frac{1}{2}} \left\{\int_0^x |f'|^2 dt\right\}^{\frac{1}{2}} \qquad (5.2.11)$$

by the Schwarz inequality. Let us write

$$A(x) = \left\{\int_0^x |f|^2 dt\right\}^{\frac{1}{2}}$$

$$B(x) = \left\{\int_0^x |f'|^2 dt\right\}^{\frac{1}{2}} .$$

Inserting (5.2.10-11) into (5.2.9) we find

$$\int_0^x |f|^2 |d\sigma| \leqq \{A^2(x) + 2A(x)B(x)\} \cdot \int_0^x |d\sigma(t)| \qquad (5.2.12)$$

for all $f \in \mathcal{D}$. For $\varepsilon > 0$,

$$\left\{\frac{1}{\sqrt{\varepsilon}} A(x) - \sqrt{\varepsilon} B(x)\right\}^2 \geqq 0 .$$

Hence

$$2A(x)B(x) \leqq \frac{1}{\varepsilon} A^2(x) + \varepsilon B^2(x) .$$

Inserting this in (5.2.12) we obtain

$$\int_0^x |f|^2 |d\sigma| \leqq \left\{\left(1 + \frac{1}{\varepsilon}\right) A^2(x) + \varepsilon B^2(x)\right\} \cdot C$$

where C is the quantity (5.2.1). Replacing ε by ε/C we find

$$\int_0^x |f|^2 |d\sigma| \leqq C(\varepsilon)A^2(x) + \varepsilon B^2(x)$$

where

$$C(\varepsilon) = 1 + \frac{C}{\varepsilon} .$$

This completes the proof.

When σ is absolutely continuous the above lemma can be found in [18, p. 339, Lemma 1] in the case when $p = 1$. Our proof appears to be simpler than the case $p = 1$ of [18]. Consequently, if we choose $\varepsilon = \frac{1}{2}$ we find that for each $f \in \mathcal{D}$,

$$\int_0^x |f|^2 |d\sigma| \leqq \frac{1}{2} \int_0^x |f'|^2 dt + C' \int_0^x |f|^2 dt \qquad (5.2.13)$$

where $C' = C\left(\frac{1}{2}\right)$.

LEMMA 5.2.3:

For every $f \in \mathcal{D}$, $f' \in L^2(0, \infty)$ and

$$\lim_{x\to\infty} f(x) = \lim_{x\to\infty} \bar{f}(x)f'(x) = 0 .$$

Proof: This can be shown as in Lemma 2 of [18]. For if

$f' \notin L^2(0, \infty)$

$$\int_0^x \{|f'|^2 dt + |f|^2 d\sigma\} \geqq \int_0^x |f'|^2 dt - \int_0^x |f|^2 |d\sigma|$$

$$\geqq \frac{1}{2} \int_0^x |f'|^2 dt - C' \int_0^x |f|^2 dt$$

by (5.2.13). Since $f \in L^2(0, \infty)$ we must have

$$\lim_{x\to\infty} \int_0^x \{|f'|^2 dt + |f|^2 d\sigma(t)\} = \infty . \qquad (5.2.14)$$

A simple calculation also shows that

$$\int_0^x (Lf)(t)\bar{f}(t)dt = \int_0^x \{|f'|^2 dt + |f|^2 d\sigma(t)\} - [\bar{f}f']_0^x . \qquad (5.2.15a)$$

However, since $f, Lf \in L^2(0, \infty)$ we must therefore have

$$\lim_{x\to\infty} \bar{f}(x)f'(x) = \infty .$$

But by taking real and imaginary parts in (5.2.15a), and noting that σ is real, the latter equation is clearly impossible. This contradiction proves that $f' \in L^2(0, \infty)$. Hence (5.2.13) implies that

$$\int_0^\infty |f|^2 |d\sigma| < \infty . \qquad (5.2.15b)$$

Thus (5.2.15a) implies that $\bar{f}(x)f'(x) \to \alpha$, as $x \to \infty$. But since $\bar{f}f' \in L(0,\infty)$, $\alpha = 0$. The following relation implies that $|f(x)|^2 \to \beta$, as $x \to \infty$:

$$|f(x)|^2 = |f(0)|^2 + \int_0^x 2\,\mathrm{re}(\bar{f}f')dt \,. \qquad (5.2.16)$$

Since $f \in L^2(0,\infty)$, $\beta = 0$. The lemma is proved.

We also obtain $|\bar{f}(x)f'(x)| \leqq \|\bar{f}f'\|_\infty$ for all large x . Thus

$$\int_0^\infty |\bar{f}(x)f'(x)||d\sigma(x)| \leqq C\|\bar{f}f'\|_\infty < \infty \qquad (5.2.17)$$

for all $f \in \mathcal{D}$. Moreover, for such f ,

$$|f'(x)|^2 = |f(0)|^2 + \int_0^x \{\bar{f}'df' + f'd\bar{f}'\}$$

$$= |f(0)|^2 + \int_0^x \{f'\bar{f} + \bar{f}'f\}d\sigma(t)$$

$$- \int_0^x \{f'\overline{Lf} - \bar{f}'Lf\}\,dt \,. \qquad (5.2.18)$$

Because of (5.2.17) the first integral in (5.2.18) is finite. Since $f', Lf \in L^2(0,\infty)$ the second integral is also finite. Thus $|f'(x)|^2 \to \gamma$ as $x \to \infty$. But since $f' \in L^2(0,\infty)$ $\gamma = 0$. Hence

$$\lim_{x\to\infty} f'(x) = 0 \qquad \text{for} \quad f \in \mathcal{D} \,. \qquad (5.2.19)$$

Since f, f' are locally of bounded variation on $[0,\infty)$

Lemma 5.2.3 and (5.2.19) imply that

$$\|f\|_\infty < \infty \quad , \quad \|f'\|_\infty < \infty \qquad f \in \mathcal{D} . \tag{5.2.20}$$

The above results essentially comprise Theorem 1 of [18] but in a more general setting.

THEOREM 5.2.3:

The operator T_0 , defined by (5.2.7), is semi-bounded from below and

$$(T_0 f , f) \geqq -C^2 (f , f) \qquad f \in \mathcal{D}(T_0) \tag{5.2.21}$$

where C is defined by (5.2.1).

Proof: Let $f \in \mathcal{D}(T_0)$. Then (5.2.16) along with an application of the Schwarz inequality shows that

$$\|f\|_\infty^2 \leqq 2 \|f\|_2 \|f'\|_2 \tag{5.2.22}$$

where the subscripts are self-explanatory. Moreover

$$\int_0^\infty |f(x)|^2 \, |d\sigma(x)| < \|f\|_\infty^2 \cdot C \tag{5.2.23}$$

Inserting (5.2.22) into the latter we obtain

$$\int_0^\infty |f(x)|^2 \, |d\sigma(x)| < 2C \|f\|_2 \|f'\|_2 \qquad f \in \mathcal{D}(T_0) . \tag{5.2.24}$$

Using (5.2.15) with $f \in \mathcal{D}(T_0)$ we have, on account of Lemma 5.2.3,

$$(T_0 f\,,\,f) = \int_0^\infty \{|f'|^2 dt + |f|^2 d\sigma(t)\}$$

$$\geqq \|f'\|_2^2 - \int_0^\infty |f|^2\,|d\sigma(t)|$$

$$\geqq \|f'\|_2^2 - 2C\|f\|_2\|f'\|_2 \qquad f \in \mathcal{D}(T_0)\ .$$

Since

$$(\|f'\|_2 - C\|f\|_2)^2 \geqq 0\ ,$$

we obtain, from the former equation, that

$$(T_0 f\,,\,f) \geqq -C^2\|f\|_2^2 = -C^2(f\,,\,f) \qquad f \in \mathcal{D}(T_0)\ .$$

This proves the theorem.

When σ is absolutely continuous and $\sigma' = q$, the lower bound becomes

$$C = \|q\|_1^2$$

so that

$$(T_0 f\,,\,f) \geqq -\|q\|_1^2(f\,,\,f) \qquad f \in \mathcal{D}(T_0)$$

The latter bound was obtained by Everitt [16, p. 146].

Consider now the recurrence relation

$$-\Delta^2 y_{n-1} + b_n y_n = \lambda y_n \qquad n = 0, 1, \ldots \tag{5.2.25}$$

where (b_n) is any real sequence. From the earlier results we know that (5.2.25) is limit-point (in the ℓ^2-sense). Consequently the results of Chapter 3 imply that the operator T_0 defined by

$$\mathcal{D}(T_0) = \{y = (y_n) \in \ell^2 : -\Delta^2 y_{n-1} + b_n y_n \in \ell^2, \; y_{-1} = 0\}$$

$$(T_0 f)_n = -\Delta^2 f_{n-1} + b_n f_n ,$$

is self-adjoint. If we suppose further that

$$C \equiv \sum_0^\infty |b_n| < \infty , \tag{5.2.26}$$

then it will follow that

$$\sum_0^\infty |b_n| |f_n|^2 < \infty \qquad f \in \mathcal{D}(T_0) \tag{5.2.27}$$

(since the $f_n \to 0$ as $n \to \infty$ for $f \in D_0$). Moreover for any sequence f_n,

$$|f_n|^2 = |f_{-1}|^2 + \sum_0^n \{f_{j-1} \Delta \bar{f}_{j-1} - \bar{f}_j \Delta f_{j-1}\}$$

where we have used partial summation.

Thus if $f \in \mathcal{D}(T_0)$, the Schwarz inequality gives

$$|f_n|^2 \leqq \left\{ \sum_0^n |f_{j-1}|^2 \right\}^{\frac{1}{2}} \left\{ \sum_0^n |\Delta f_{j-1}|^2 \right\}^{\frac{1}{2}}$$

$$+ \left\{ \sum_0 |f_j|^2 \right\}^{\frac{1}{2}} \left\{ \sum_0^n |\Delta f_{j-1}|^2 \right\}^{\frac{1}{2}}$$

and so

$$\sup_n |f_n|^2 \leqq 2\|f\|_2 \|\Delta f\|_2 \qquad f \in \mathcal{D}(T_0) \tag{5.2.28}$$

where

$$\|f\|_2 = \left\{ \sum_0^\infty |f_n|^2 \right\}^{\frac{1}{2}}$$

and

$$\|\Delta f\|_2 = \left\{ \sum_0^\infty |\Delta f_{n-1}|^2 \right\}^{\frac{1}{2}} .$$

Inserting (5.2.28) into (5.2.27) we find

$$\sum_0^\infty |b_n||f_n|^2 \leqq 2\|f\|_2 \|\Delta f\|_2 \cdot C \qquad f \in \mathcal{D}(T_0) . \tag{5.2.29}$$

Now, from (4.1.7) we see that, for $f \in \mathcal{D}(T_0)$,

$$(T_0 f , f) = \sum_0^\infty \{ |\Delta f_{n-1}|^2 + b_n |f_n|^2 \}$$

$$\geqq \sum_0^\infty |\Delta f_{n-1}|^2 - \sum_0^\infty |b_n||f_n|^2$$

$$> \|\Delta f\|_2^2 - 2\|f\|_2 \|\Delta f\|_2 C$$

$$\geqq -C^2 \|f\|_2 .$$

Thus

$$(T_0 f , f) \geqq -C^2(f , f) \qquad f \in \mathcal{D}(T_0)$$

consequently we obtain the analogous bound of Theorem 5.2.3 in the case when $\sigma(t)$ is given by

$$\sigma(t_n) - \sigma(t_n - 0) = -b_n ,$$

since then

$$\int_0^\infty |d\sigma(t)| = \sum_0^\infty |b_n| .$$

The above is the discrete analog of Everitt's theorem [16].

APPENDIX I

§1. FUNCTIONS OF BOUNDED VARIATION:

We summarize here those definitions and results concerning functions of bounded variation which were used in the preceding chapters. Most of these results are well-known and can be found in [1, p. 127ff.] thus we omit the proofs.

DEFINITION 1.1:

A function $\sigma(x)$ defined over a finite real interval (a, b) is of *bounded variation over* (a, b) if the set of values of the sum

$$\sum_{r=0}^{n-1} |\sigma(x_{r+1}) - \sigma(x_r)| \tag{I.1.1}$$

is bounded for all n and all possible subdivisions $\{x_r\}$ of (a, b), $a = x_0 < x_1 < \cdots < x_n = b$. The least common upper bound of all such sums is called the total variation of σ over (a, b) and is sometimes denoted by $\mathrm{Var}\{\sigma(x) ;\ a, b\}$ or $V_f(a, b)$.

THEOREM A:

If $\sigma(t)$ is of bounded variation on (a, b) then

a) $\sigma(x+0)$, $\sigma(x-0)$ exist for each $x \in (a, b)$.

b) $\sigma(b-0)$, $\sigma(a+0)$ both exist.

c) The points of discontinuity form a finite or at most denumerable set.

d) σ is bounded on (a, b) .

If a is finite and $b = \infty$ we say that σ is of bounded variation over (a, ∞) if $\sigma(\infty)$ is defined and that the requirement of uniform boundedness of (I.1.1) be satisfied when $b = x_n = \infty$.

In this work we shall assume further that

$$\sigma(\infty) = \lim_{t\to\infty} \sigma(t) \tag{I.1.2}$$

exists so that $\sigma(t)$ cannot have a jump at infinity.

<u>DEFINITION 1.2</u>:

A real-valued function f defined on $[a, b]$ is *absolutely continuous* on $[a, b]$ if for every $\varepsilon > 0$ there is a $\delta > 0$ such that

$$\sum_{k=1}^{n} |f(x_k + h_k) - f(x_k)| < \varepsilon \tag{I.1.3}$$

for every n disjoint subintervals $(x_k, x_k + h_k)$ of $[a, b]$ such that

$$\sum_{k=1}^{n} h_k < \delta . \qquad (I.1.4)$$

THEOREM B:

Let f be an absolutely continuous function on $[a, b]$.

Then a) f is continuous.

b) f has a derivative almost everywhere.

c) f is of bounded variation on (a, b).

d) f' is Lebesgue integrable on (a, b) and

$$\int_a^b f'(x)dx = f(b) - f(a) .$$

For further results concerning functions of bounded variation and the Lebesgue theory of integration we refer to [1].

§2. THE RIEMANN-STIELTJES INTEGRAL:

DEFINITION 2.1:

Let f, σ be real valued functions defined on some finite interval $[a, b]$.

For any partition $\{x_r\}$ of (a, b),

$$a = x_0 < x_1 < \cdots < x_n = b \qquad (I.2.1)$$

we form the sum

$$S = \sum_{r=0}^{n-1} f(\xi_r)\left(\sigma(x_{r+1}) - \sigma(x_r)\right) \qquad (I.2.2)$$

where $\xi_r \in [x_r , x_{r+1}]$.

If, as $n \to \infty$ and $\max|x_{r+1} - x_r| \to 0$ the sum S tends to a unique limit for all partitions $\{x_r\}$ and for all choices of $\xi_r \in [x_r , x_{r+1}]$ the limit is called the *Stieltjes integral of* f *with respect to* σ written

$$\int_a^b f(x)d\sigma(x) \; . \qquad \text{(I.2.3)}$$

If the "distribution function" σ is a step-function the integral reduces to a sum [1, p. 148].

THEOREM C:

If f is continuous on $[a , b]$ and σ is of bounded variation over (a , b) the integral (I.2.3) exists [1, p. 159].

THEOREM D:

If f is continuous on $[a , b]$ and σ is of bounded variation over (a , b) then

$$\left| \int_a^b f(x)d\sigma(x) \right| \leq \sup_{[a,b]} |f(x)| \cdot \mathrm{Var}\{\sigma(x) : a , b\} .$$

Integrals over infinite intervals are understood in the improper sense so that

$$\int_a^\infty f(x)d\sigma(x) = \lim_{b\to\infty} \int_a^b f(x)d\sigma(x) \qquad \text{(I.2.4)}$$

whenever the limit exists.

THEOREM E:

If f is continuous for all real finite x uniformly bounded on $[a, \infty)$ and σ is of bounded variation over $[a, \infty)$ then the integral

$$\int_a^\infty f(x)\,d\sigma(x) \qquad (I.2.5)$$

exists [3, p. 422].

THEOREM F:

If f is continuous on $[a, b]$ and σ is of bounded variation over $[a, b]$ then [1, p. 144]

$$\int_a^b f(x)\,d\sigma(x) = f(b)\sigma(b) - f(a)\sigma(a) - \int_a^b \sigma(x)\,df(x) .$$

THEOREM G:

If f, g are continuous on $[a, b]$ and σ is of bounded variation over $[a, b]$ then

a) $F(x) = \int_a^x f(s)\,d\sigma(s)$ is of bounded variation over $[a, b]$.

b)
$$\int_a^b g(s)\,dF(s) = \int_a^b gf\,d\sigma .$$

c) If σ is continuous (right-continuous) at x then $F(x)$ is continuous (right-continuous) at x .

d) If for some $x_0 \in (a, b)$ $f(x_0) = 0$, then $F(x)$ is continuous at x_0 .

e) $$\mathrm{Var}\{\sigma(x) : a, b\} = \int_a^b |d\sigma(x)|$$

where the integral is interpreted as the limit of (I.1.1) as $n \to \infty$ the subdivisions being increasingly fine so that $\max|x_{r+1} - x_r| \to 0$ as $r \to \infty$.

Proof: a), b), c) can be found in [1, p. 161 - 62].

d) follows from the relation

$$F(x_0 + 0) - F(x_0 - 0) = \int_{x_0-0}^{x_0+0} f(s)d\sigma(s)$$

$$= f(x_0)\big(\sigma(x_0 + 0) - \sigma(x_0 - 0)\big)$$

e) follows from the definition.

THEOREM H:

Let f be a real-valued right-continuous function on $[a, b]$.

Then

$$\lim_{h\to 0+} \frac{1}{h} \int_x^{x+h} f(s)ds = f(x) \tag{I.2.6}$$

for all $x \in [a, b)$.

Proof: The right-continuity of f implies that given $\varepsilon > 0$, there is a $\delta > 0$ such that

$$|f(s) - f(x)| < \varepsilon \tag{I.2.7}$$

whenever $x \leq s \leq x + \delta$, $x \in [a, b)$.

Let $x \in [a, b)$, $\varepsilon > 0$.

$$\left| \frac{1}{h} \int_x^{x+h} f(s)ds - f(x) \right| = \left| \frac{1}{h} \int_x^{x+h} \big(f(s) - f(x)\big)ds \right|$$

$$\leqq \frac{1}{h} \int_x^{x+h} |f(s) - f(x)|ds .$$

We restrict h so that $0 < h < \delta$ where the δ is the same as that in (I.2.7). Using now (I.2.7) where $0 < h < \delta$ we find that

$$\left| \frac{1}{h} \int_x^{x+h} f(s)ds - f(x) \right| \leqq \frac{1}{h} \int_x^{x+h} |f(s) - f(x)|ds$$

$$\leqq \frac{1}{h} \cdot \varepsilon \cdot h = \varepsilon$$

whenever $0 < h < \delta$. This completes the proof.

THEOREM J:

Let f be absolutely continuous on $[a, b]$ and suppose that f has a right-derivative $f'_+(x)$ at each point $x \in [a, b]$. Let g be a right-continuous function of bounded variation over $[a, b]$ then

$$\lim_{h\to 0+} \frac{1}{h} \int_x^{x+h} g(s)\,df(s) = g(x) f'_+(x) \qquad (I.2.8)$$

for each $x \in [a, b)$.

Proof: The integral above exists for h sufficiently small and $x \in [a, b)$ by virtue of Theorem F. It now suffices to prove the theorem when f is increasing (since every function of bounded variation is the difference of two increasing functions).

Let $x \in [a, b)$ and assume f is increasing. Then [1, p. 160, Theorem 7.30] implies the existence of a number c where

$$\inf_{s\in[x,x+h]} g(s) \leqq c \leqq \sup_{s\in[x,x+h]} g(s) \qquad (I.2.9)$$

with the property that

$$\int_x^{x+h} g(s)\,df(s) = c\big(f(x+h) - f(x)\big) . \qquad (I.2.10)$$

Dividing (I.2.10) by h , we see from (I.2.9) and the right-continuity of g that $c \to g(x)$ as $h \to 0+$.

Hence

$$\lim_{h\to 0+} \frac{1}{h} \int_x^{x+h} g(s)\,df(s) = g(x) \lim_{h\to 0+} \frac{f(x+h) - f(x)}{h}$$

$$= g(x) f'_+(x)$$

for $x \in [a, b)$. The result now follows.

THEOREM K:

Let f, g satisfy the hypotheses of Theorem J and f' be Riemann integrable. Then

$$\int_a^t g(s)df(s) = \int_a^t g(s)f_+'(s)ds \ . \qquad (I.2.11)$$

Proof: Let $A(t)$, $B(t)$ be defined by the left side and right side of (I.2.11) respectively. Since gf_+' is integrable $B(t)$ is continuous and since f is absolutely continuous $A(t)$ is continuous. Apart from the jumps of g, the derivative appearing in (I.2.11) is a two-sided derivative in which case (I.2.11) holds for almost all t. On the other hand if t is a jump point of g then since the continuity of f at t implies that of A at t, $A(t)$ must equal $B(t)$ at the jump points of g and so $A(t) = B(t)$ everywhere on $[a, b)$.

§3. GENERAL THEORY OF VOLTERRA-STIELTJES INTEGRAL EQUATIONS:

In this section we shall summarize the basic tools required for the development of the theory of integral equations of the type (1.0.0). In the case when $p(t) = 1$ the theory was developed in [3, p. 339]. The case of the general $p(t)$ is not very different from the case when $p(t) = 1$ and so we shall only state those results whose proofs would be similar to those in [3] with the appropriate modifications. We shall begin by proving an existence and uniqueness result.

THEOREM I.3.1: (for a related result see [79])

Let $\sigma(t)$ be right-continuous and of bounded variation over $[a, b]$ and let $p(t)$ be a positive right-continuous function of bounded variation on $[a, b]$ with the property that

$$\frac{1}{p(t)} \in L(a,b). \tag{I.3.1}$$

Then the integral equation

$$y(t) = \alpha + \beta \int_a^t \frac{1}{p} + \int_a^t \frac{1}{p(s)} \int_a^s y\, d\sigma\, ds \tag{I.3.2}$$

has a unique continuous solution on $[a, b]$ for given α, β.

Proof: We shall essentially use the Picard method of successive approximations.

(I.3.1) implies that the function $y_0(t)$ defined by

$$y_0(t) = \alpha + \beta \int_a^t \frac{1}{p} \tag{I.3.3}$$

is absolutely continuous on $[a, c]$ where $c < b$ and so

$$M = \sup_{[a,c]} |y_0(t)| \tag{I.3.4}$$

exists and is finite.

We now define y_n by recurrence on n.

$$y_{n+1}(t) = \int_a^t \frac{1}{p(s)} \int_a^s y_n \, d\sigma \, ds \tag{I.3.5}$$

for $n = 0, 1, 2, \ldots, t \in [a, c]$.

Let V be defined by

$$V = \int_a^c |d\sigma(s)| \tag{I.3.6}$$

i.e. V is the total variation of σ over $[a, c]$. We claim that

$$|y_n(t)| \leqq \frac{MV^n}{n!} \left\{ \int_a^t \frac{1}{p} \right\}^n \quad t \in [a, c] \tag{I.3.7}$$

for $n = 0, 1, \ldots$. We prove this by induction on n . (I.3.7) holds if $n = 0$. If we assume that (I.3.7) is true for $n = m$, then

$$|y_{m+1}(t)| \leqq \int_a^t \frac{1}{p(s)} \int_a^s |y_m(x)| \, |d\sigma(x)| \, ds \tag{I.3.8}$$

$$\leqq \int_a^t \frac{1}{p(s)} \int_a^s \left\{ \frac{MV^m}{m!} \left(\int_a^x \frac{1}{p} \right)^m \right\} |d\sigma(x)| \, ds \tag{I.3.9a}$$

$$\leqq \int_a^t \frac{1}{p(s)} \left\{ \frac{MV^m}{m!} \left(\int_a^s \frac{1}{p} \right)^m \right\} \cdot \int_a^s |d\sigma(x)| \, ds$$

$$\leqq \frac{MV^m}{m!} \cdot V \int_a^t \frac{1}{p(s)} \left(\int_a^s \frac{1}{p} \right)^m ds$$

$$= \frac{MV^{m+1}}{(m+1)!} \left(\int_a^t \frac{1}{p} \right)^{m+1} \tag{I.3.9b}$$

where (I.3.9) is (I.3.7) with $n = m+1$. Hence (I.3.7) holds for all n. Consequently the series

$$\sum_{n=0}^{\infty} y_n(t) = y(t) \tag{I.3.10}$$

is uniformly convergent on $[a, c]$ and

$$y_0(t) + \int_a^t \frac{1}{p(s)} \int_a^s y(x)\,d\sigma(x)\,ds$$

$$= y_0(t) + \int_a^t \frac{1}{p(s)} \int_a^s \sum_{n=0}^{\infty} y_n(x)\,d\sigma(x)\,ds$$

$$= y_0(t) + \sum_{n=0}^{\infty} \int_a^t \frac{1}{p(s)} \int_a^s y_n(x)\,d\sigma(x)\,ds$$

$$= y_0(t) + \sum_{n=0}^{\infty} y_{n+1}(t)$$

$$= y(t)$$

where the interchange of the order of integration and summation is justified by the uniform convergence of the series, [1, p. 225]. So (I.3.10) satisfies the integral equation (I.3.2) and hence represents a solution. From (I.3.5) we see that each y_n is continuous, essentially because y_0 is, and therefore the limit function $y(t)$ must also be continuous because the convergence is uniform.

To show that $y(t)$ is unique we use the extended Gronwall lemma [3, p. 455], which states that if $\rho(x)$,

$x \in [a, b]$, is non-negative and continuous and σ is non-negative and continuous and σ is non-decreasing and right-continuous, and for $a \le x \le b$,

$$\rho(x) \leqq c_0 + c_1 \int_a^x \rho(t)d\sigma(t) \qquad (I.3.11)$$

where $c_0 > 0$, $c_1 > 0$ then

$$\rho(x) \leqq c_0 \exp\{c_1[\sigma(x) - \sigma(a)]\} . \qquad (I.3.12)$$

Assume if possible that $y(t)$, $z(t)$ are two solutions to (I.3.2). Then

$$|y(t) - z(t)| \leqq \int_a^t \frac{1}{p(s)} \int_a^s |y(x) - z(x)||d\sigma(x)|ds$$

$$\leqq \left(\int_a^t \frac{1}{p} \right) \int_a^t |y(x) - z(x)||d\sigma(x)|$$

$$\leqq \left(\int_a^c \frac{1}{p} \right) \int_a^t |y(x) - z(x)||d\sigma(x)| . \qquad (I.3.13)$$

The extended Gronwall lemma (I.3.12) now implies that $|y(t) - z(t)| = 0$ for $t \in [a, c]$ and so $y(t) = z(t)$ on $[a, c]$ for any $c < b$. Hence $y(t) = z(t)$ everywhere on $[a, b]$.

<u>REMARK</u>:

It is possible to show Theorem I.3.1 by a transformation

of the independent variable. For if $\frac{1}{p} \in L(a, b)$ then the transformation

$$t \mapsto \tau(t) = \int_a^t \frac{1}{p} \tag{I.3.14}$$

will transform (1.0.0) into an equation of the form

$$y'(\tau) = c_1 + \int_0^\tau y(s)d\sigma(s) \qquad \tau \in [0, \tau(b)] \tag{I.3.15}$$

where $(') = \frac{d}{d\tau}$ [1, pp. 144 - 45]. The existence and uniqueness results for (I.3.15) are discussed in [3, p. 341].

THEOREM I.3.2:

The solution (I.3.2) has a right-derivative in $[a, b]$ satisfying the integro-differential equation

$$p(t)y'(t) = \beta + \int_a^t y(s)d\sigma(s) \tag{I.3.16}$$

where $\beta = p(a)y'(a)$, $t \in [a, b]$.

If both σ, p are continuous at t or if $y(t) = 0$ then $y'(t)$ is an ordinary derivative.

Proof: The proof is similar, with the appropriate changes to that in [3, p. 346] and so is omitted.

Other theorems dealing with equations of the form (I.3.15) can be found in Chapter 11 of [3].

THEOREM I.3.3:

There exists two linearly independent solutions $y(t)$, $z(t)$ of (I.3.2) such that

$$p(t)\big(y(t)z'(t) - z(t)y'(t)\big) = 1 . \qquad \text{(I.3.17)}$$

where the "derivatives" are generally right-derivatives [3, p. 348]. In fact the Wronskian (I.3.17) is continuous at $t = b$.

We shall write the inhomogeneous equation associated with (I.3.2) as

$$y(t) = \alpha + \beta \int_a^t \frac{1}{p} + \int_a^t \frac{1}{p(s)} \int_a^s y d\sigma\, ds + \int_a^t \frac{f(s)}{p(s)}\, ds$$

(I.3.18)

where f is a right-continuous function of bounded variation on $[a, b]$.

THEOREM I.3.4:

Let σ, f, p be right-continuous functions of bounded variation and $p(t) > 0$, $t \in [a, b]$. Then the function

$$\psi(t) = y(t) \int_a^t z(s)df(s) - z(t) \int_a^t y(s)df(s) \qquad \text{(I.3.19)}$$

is a solution of (I.3.18) for $\alpha = 0$, $\beta = -f(a)$.

Here y, z are two linearly independent solutions of

$$p(t)y'(t) = 1 + \int_a^t y d\sigma \tag{I.3.20}$$

$$p(t)z'(t) = \int_a^t z d\sigma \tag{I.3.21}$$

chosen so that $p(t)\big(y'(t)z(t) - z'(t)y(t)\big) = 1$ for all $t \in [a, b]$.

Proof: The proof could be carried out as in [3, p. 351] with the appropriate modifications or by using differentials. For ψ, as given by (I.3.19), has a right-derivative ψ' given by [3, p. 352]

$$\psi'(t) = y'(t)\int_a^t z df - z'(t)\int_a^t y df \tag{I.3.22}$$

and so multiplying (I.3.22) by $p(t)$ we see that $p(t)\psi'(t)$ is of bounded variation over $[a, b]$ so that

$$d\big(p(t)\psi'(t)\big) = \left(\int_a^t z df\right) d\big(p(t)y'(t)\big) - \left(\int_a^t y df\right) d\big(p(t)z'(t)\big)$$
$$+ p(t)y'(t)d\left(\int_a^t z df\right) - p(t)z'(t)d\left(\int_a^t y df\right) \tag{I.3.23}$$

$$= y(t)\left(\int_a^t z df\right) d\sigma(t) - z(t)\left(\int_a^t y df\right) d\sigma(t)$$
$$+ p(t)y'(t)z(t)df(t) - p(t)z'(t)y(t)df(t) \tag{I.3.24}$$

$$= \left\{y(t) \int_a^t z\,df - z(t) \int_a^t y\,df\right\} d\sigma(t)$$

$$+ p(t)\{y'(t)z(t) - y(t)z'(t)\}df(t)$$

$$= \psi(t)d\sigma(t) + df(t) \qquad \text{(I.3.25)}$$

on account of (I.3.19) and the constancy of the Wronskian. That ψ is continuous can almost be read out from (I.3.19). For

$$\psi(t+0) - \psi(t-0) = y(t) \int_{t-0}^{t+0} z\,df - z(t) \int_{t-0}^{t+0} y\,df$$

$$= y(t)z(t)\big(f(t+0) - f(t-0)\big)$$

$$- z(t)y(t)\big(f(t+0) - f(t-0)\big)$$

$$= 0$$

so that ψ is continuous at t and thus everywhere.

Thus we can integrate (I.3.25) over $[a, t]$ to get

$$p(t)\psi'(t) = p(a)\psi'(a) + \int_a^t \psi(s)d\sigma(s) + f(t) - f(a)$$

or dividing throughout by $p(t)$ and integrating again over $[a, t]$ we obtain

$$\psi(t) = \alpha + \big(\beta - f(a)\big) \int_a^t \frac{1}{p} + \int_a^t \frac{1}{p} \int_a^s \psi d\sigma ds + \int_a^t \frac{f}{p}$$

where $\alpha = \psi(a)$, $\beta = p(a)\psi'(a)$. But $\psi(a) = p(a)\psi'(a) = 0$

thus we find that the latter equation is equivalent to (I.3.18) with $\alpha = 0$, $\beta = -f(a)$.

REMARK:

If $h(t)$ is the solution of the homogeneous equation

$$p(t)h'(t) = c + \int_a^t h(s)d\sigma(s) \qquad (I.3.26)$$

satisfying the initial conditions

$$h(a) = \alpha \qquad (I.3.27)$$

$$p(a)h'(a) = \beta + f(a) \qquad (I.3.28)$$

then the solution ψ of (I.3.18) is given by

$$\psi(t) = h(t) + y(t)\int_a^t zdf - z(t)\int_a^t ydf \qquad (I.3.29)$$

i.e. (I.3.29) is the solution of (I.3.18) corresponding to the initial conditions $\psi(a) = \alpha$, $p(a)\psi'(a) = \beta$.

§4. CONSTRUCTION OF THE GREEN'S FUNCTION

We begin by defining two linear forms U_1 , U_2 by

$$U_i y = \sum_{j=1}^{2} \left\{M_{ij}\, y^{(j-1)}(a) + N_{ij}\, p(b)y^{(j-1)}(b)\right\} \qquad (I.4.1)$$

for $i = 1, 2$, where y is a solution of some Stieltjes integral equation (I.3.0) or (I.3.2). The derivative appearing in (I.4.1) will, in general, be taken to mean a right-derivative and the M, N are constants.

By

$$U(y) = 0 \tag{I.4.2}$$

we shall mean both

$$\left.\begin{aligned} U_1(y) &= 0 \\ U_2(y) &= 0 \,. \end{aligned}\right. \qquad \text{(I.4.3-4)}$$

We assume at the outset that the boundary conditions (I.4.3-4) are linearly independent. If the homogeneous problem

$$\left.\begin{aligned} y(t) &= \alpha + \beta \int_a^t \frac{1}{p} + \int_a^t \frac{1}{p(s)} \int_a^s y d\sigma \, ds \\ U(y) &= 0 \end{aligned}\right\} \qquad \text{(I.4.5)}$$

has only the zero solution then we say that the homogeneous equation with homogeneous boundary conditions (I.4.3-4) is *incompatible* [5, p. 73]. It is *compatible* if a nontrivial solution satisfies (I.4.5).

Let y, z be two linearly independent solutions of (I.3.2) such that $y(a) = 0$, $p(a)y'(a) = 1$; $z(a) = 1$, $p(a)z'(a) = 0$. Then

$$p(t)\{y'(t)z(t) - z'(t)y(t)\} = 1 \tag{I.4.6}$$

for all t.

We now define a new function $K(x, t)$ by

$$K(x, t) = \begin{cases} 0 & \text{if } x < t \\ y(x)z(t) - z(x)y(t) & \text{if } x \geqq t \end{cases} \tag{I.4.7}$$

where $a \le x \le b$ and $a < t < b$.

From (I.4.7) we see that, for fixed x, $K(x, t)$ is a continuous function of t and similarly of x for fixed t. A simple computation shows that for $a < t < b$,

$$K_x(t+0, t) - K_x(t-0, t) = \frac{1}{p(t)} \tag{I.4.8}$$

where $K_x(t+0, t)$ represents the right-derivative of K with respect to x evaluated at t.

Moreover from Theorem I.3.4 the function $\phi(x)$ defined by

$$\phi(x) = \int_a^b K(x, t)df(t) \tag{I.4.9}$$

satisfies the inhomogeneous problem (I.3.18) but not the boundary conditions (I.4.3-4). It is therefore necessary to modify the function K so that, as a function of x, it should satisfy the boundary conditions.

Therefore let

$$G(x\,,\,t) = \alpha_1 y(x) + \alpha_2 z(x) + K(x\,,\,t) \qquad \text{(I.4.10)}$$

where we choose the α_i so that, for fixed $t \in (a\,,\,b)$, G as a function of x satisfies $UG = 0$.

i.e.

$$U_i(G) = U_i(K) + \alpha_1 U_i(y) + \alpha_2 U_i(z)$$

$$= 0 \qquad i = 1\,,\,2\;.$$

Since $U_i(K)$ can be made continuous on $[a\,,\,b]$ we obtain the following system of equations

$$\begin{bmatrix} U_1(y) & U_1(z) \\ U_2(y) & U_2(z) \end{bmatrix} \begin{bmatrix} \alpha_1 \\ \alpha_2 \end{bmatrix} = \begin{bmatrix} -U_1(K) \\ -U_2(K) \end{bmatrix} \qquad \text{(I.4.11)}$$

(I.4.11) can be solved uniquely for each $t \in [a\,,\,b]$ if the determinant of the matrix is different from zero for such t . In such a case the resulting solutions $\alpha_i(t)$ will be continuous in t for $t \in [a\,,\,b]$.

We now show that (I.4.5) is compatible if and only if the determinant of the matrix appearing in (I.4.11) is zero. For the equation is compatible if and only if (I.4.5) admits a non-trivial solution ω which is the case if and only if there exists c_1 , c_2 not both zero such that

$$\omega(t) = c_1 y(t) + c_2 z(t)$$

and

$$U_1(\omega) = U_2(\omega) = 0 .$$

The latter is true if and only if

$$\begin{aligned} (U_1(\omega) =) \qquad & c_1 U_1(y) + c_2 U_1(z) = 0 \\ (U_2(\omega) =) \qquad & c_1 U_2(y) + c_2 U_2(z) = 0 \end{aligned}$$

admits nontrivial solutions and this is equivalent to the determinant of the matrix appearing (I.4.11) being zero. Thus since we assumed that (I.4.5) was incompatible the latter determinant must be different from zero and so (I.4.11) admits unique solutions $\alpha_i(t)$ for each t. This said, the function $G(x, t)$ defined in (I.4.10) will satisfy $UG = 0$ as a function of x and is called the *Green's function* for the problem (I.3.18), (I.4.3-4).

<u>THEOREM I.4.1</u>:

Whenever (I.4.5) is incompatible there exists a unique function $G(x, t)$ defined for $a \le x , t \le b$ having the following properties:

a) $G(x, t)$ is continuous in x, t jointly and absolutely continuous in x for fixed $t \in [a, b]$.

b) As a function of x, G satisfies (I.4.5) except when $x = t$ and the boundary conditions $UG = 0$ for $a \le t \le b$.

c) The solution of the inhomogeneous problem (I.3.18), (I.4.3-4) is given by

$$y(x) = \int_a^b G(x, t)\,df(t) . \tag{I.4.12}$$

d) When $x = t$, $G_x(x, t)$ has a jump of magnitude

$$G_x(t+0, t) - G_x(t-0, t) = \frac{1}{p(t)} + \alpha_1(t)\{y'(t) - y'(t-0)\} + \alpha_2(t)\{z'(t) - z'(t-0)\} \tag{I.4.13}$$

where α_1, α_2, y, z were defined earlier.

REMARKS:

1. If, in (I.3.2), we had that p, σ were continuous on $[a, b]$ then, as we saw in section 1 of this Appendix, the solutions would have continuous derivatives and therefore the "extra" terms in (I.4.13) would disappear leaving

$$G_x(t+0, t) - G_x(t-0, t) = \frac{1}{p(t)} . \tag{I.4.14}$$

(I.4.14) is, in particular, satisfied when σ, p are both $C'(a, b)$ in which case the Green's function is that of an ordinary differential equation [9, p. 192].

2. If x is a point at which either σ or p is discontinuous the derivative of the Green's function will then also have a jump there of magnitude

$$G_x(x+0, t) - G_x(x-0, t) = \{\alpha_1(t) - z(t)\}\{y'(x) - y'(x-0)\} + \{\alpha_2(t) + z(t)\}\{z'(x) - z'(x-0)\} . \tag{I.4.15}$$

This is to be expected at a point of discontinuity of either σ or p since then the solutions may have a discontinuous derivative at that point and this would affect the Green's function.

Again we note that when both σ, p are continuous, in addition to the usual hypotheses, the Green's function has a derivative (see (I.4.15)).

APPENDIX II

§1. COMPACTNESS IN L^p AND OTHER SPACES:

In this Appendix we prove certain theorems which pertain to Chapter 2 and state, without proofs, certain fundamental theorems.

We shall use the following version of Schauder's Fixed Point Theorem.

THEOREM II.1.1:

Let X be a convex subset of a Banach space and A be a continuous map leaving X invariant, i.e. $AX \subset X$, and such that AX is compact. Then A has a fixed point [57, p. 25].

THEOREM II.1.2: (Riesz [54, p. 137]

A family F of functions in $L^p(-\infty, \infty)$, $p \geq 1$, is compact if and only if

a) There is an $M > 0$ such that, for all $f \in F$,

$$\|f\|_p \leq M \tag{II.1.1}$$

b) For any $\varepsilon > 0$ there is a $\delta(\varepsilon) > 0$ such that, for all $f \in F$,

$$\|f(x+h) - f(x)\|_p < \varepsilon \qquad \text{(II.1.2)}$$

whenever $|h| < \delta$.

c) If $E_A = \{x \in \mathbb{R} : |x - x_0| > A \text{ , } x_0 \text{ fixed}\}$ then, for all $f \in F$,

$$\lim_{A\to\infty} \|f\|_{E_A} = 0 . \qquad \text{(II.1.3)}$$

where the norm is the induced L^p-norm on E_A .

The above theorem is the "L^p-analog" of the Arzelà-Ascoli theorem.

COROLLARY II.1.2:

Let F be a family of functions in $L^p[a, \infty)$, $p \geq 1$, $a > -\infty$, satisfying the following conditions:

a) There is an $M > 0$ such that, for all $f \in F$,

$$\|f\|_p \leq M . \qquad \text{(II.1.4)}$$

b) If $E_A = \{x : A \leq x < \infty\}$ then, for given $\varepsilon > 0$, and for all $f \in F$,

$$\|f\|_{E_A} \leq \varepsilon \qquad \text{(II.1.5)}$$

if A is sufficiently large.

c) For $\varepsilon > 0$ there is a $\delta(\varepsilon) > 0$ such that, for all $f \in F$,

$$\left| \int_a^{a+h} |f(x)|^p dx \right| < \varepsilon \qquad \text{(II.1.6)}$$

whenever $|h| < \delta$.

d) For $\varepsilon > 0$ there is an $\eta(\varepsilon) > 0$ such that, for all $f \in F$,

$$\|f(x+h) - f(x)\|_p < \varepsilon \qquad \text{(II.1.7)}$$

whenever $|h| < \eta$.

Then F is compact (in the $L^p[a, \infty)$-sense).

<u>Proof:</u> Set $F = 0$ on $(-\infty, a)$ i.e. If $f \in F$ then $f = 0$ a.e. on $(-\infty, a)$. Then F is a family of functions in $L^p(\mathbb{R})$ to which we shall apply Theorem II.1.2.

i) (II.1.1) is clearly verified on account of (II.1.4) and $F = 0$ on $(-\infty, a)$.

ii) For all $f \in F$, $\varepsilon > 0$, there is an $A_0(\varepsilon) > 0$ such that

$$\int_A^{\infty} |f(x)|^p dx < \varepsilon \qquad A \geqq A_0$$

and clearly

$$\int_{-\infty}^{-A} |f(x)|^p dx = 0 < \varepsilon \qquad A \geqq A_0 .$$

This implies that (II.1.3) is satisfied.

iii) Let $\varepsilon > 0$. Suppose $|h| < \delta^* = \min\{\delta(\varepsilon/2), \eta(\varepsilon/2)\}$.

If $0 < h < \delta$, $f \in F$,

$$\int_{-\infty}^{\infty} |f(x+h) - f(x)|^p \, dx = \left\{ \int_{-\infty}^{a} + \int_{a}^{\infty} \right\} |f(x+h) - f(x)|^p \, dx$$

$$= \int_{-\infty}^{a} |f(x+h)|^p \, dx + \int_{a}^{\infty} |f(x+h) - f(x)|^p \, dx$$

$$= \int_{a-h}^{a} |f(x+h)|^p \, dx + \int_{a}^{\infty} |f(x+h) - f(x)|^p \, dx$$

$$= \int_{a}^{a+h} |f(t)|^p \, dt + \int_{a}^{\infty} |f(x+h) - f(x)|^p \, dx$$

$$< \varepsilon/2 + \varepsilon/2 = \varepsilon$$

on account of (II.1.6-7).

If $-\delta < h < 0$, $f(x+h) - f(x) = 0$ a.e. on $(-\infty, a)$ so that

$$\int_{\infty}^{\infty} |f(x+h) - f(x)|^p \, dx = 0 + \int_{a}^{\infty} |f(x+h) - f(x)|^p \, dx$$

$$< \varepsilon/2$$

$$< \varepsilon$$

for such h .

Thus if $|h| < \delta^*$

$$\int_{-\infty}^{\infty} |f(x+h) - f(x)|^p \, dx < \varepsilon$$

for all $f \in F$ which verifies (II.1.2). Consequently

Theorem II.1.1 implies that F is compact in the $L^p(\mathbb{R})$-sense. That is, given any sequence $(f_n) \in F$ there is a subsequence, which we rewrite as f_n, which converges to an element $f \in F$.

i.e.

$$\int_{-\infty}^{\infty} |f_n - f|^p \, dx \to 0 \qquad n \to \infty .$$

But since $f_n = 0$ a.e. on $(-\infty, a)$ the latter equation implies that $f = 0$ a.e. on $(-\infty, a)$ and so

$$\int_{a}^{\infty} |f_n - f|^p \, dx \to 0 \qquad n \to \infty$$

which says that $f_n \to f$ in the $L^p[a, \infty)$-sense and so F is compact in this sense. This completes the proof.

LEMMA II.1.1:

In the proof of Theorem 2.3.1, B_n is compact.

Proof: Let K be a compact set in $[T, T+n)$ for fixed n. Let (t_i) be an enumeration of the rationals in K and let x_k be an arbitrary sequence in B_n. We write z_k for x_k' (the sequence of right-derivatives of x_k). We will show that x_k has a subsequence which converges uniformly on K.

Since $(x_k) \subset B_n$ the z_k are uniformly bounded and hence there is a subsequence $z_{1,n}(t)$ which converges at t_1. Similarly there is a subsequence $z_{2,n}$ of $z_{1,n}$ which converges at t_2 and at t_1 and so on. Continuing in this way there is a subsequence $z_{j,n}$ of $z_{j-1,n}$ which converges at $t_1, t_2, \ldots, t_{j-1}$ and at $t = t_j$. Moreover $z_{k,n}$ is a subsequence of $z_{j,n}$ if $k \geq j$. The diagonal sequence $z_{m,m}$ then converges at each rational (since $\{z_{m,m} : m \geq k\}$ is a subsequence of $\{z_{k,m} : m \geq 1\}$ and so converges at $t = t_k$).

Rewriting z_m for $z_{m,m}$ we define a function $z(t)$ by

$$z(t_i) = \lim_{m\to\infty} z_m(t_i) \qquad i = 1, 2, \ldots . \qquad \text{(II.1.8)}$$

This limit exists from the above considerations. The domain of definition of $z(t)$ can be extended to the irrationals in K by right-continuity. For if t is irrational let s_n be a decreasing sequence of rationals converging to t. Then $z(s_n)$ is uniformly bounded and so there is a subsequence $s_n \downarrow t$ for which the limit of $z(s_n)$ exists as $n \to \infty$.

For such t we let

$$z(t) = \lim_{n\to\infty} z(s_n) . \qquad \text{(II.1.9)}$$

This is well-defined for if r_n is another rational sequence decreasing to t for which there is a subsequence r_n such that $\lim z(r_n)$ exists as $n \to \infty$, then, denoting the latter limit by $z^*(t)$,

$$|z(t) - z^*(t)| = \lim_{n\to\infty} |z(s_n) - z(r_n)|$$

$$\leq \lim_{n\to\infty} \left\{ a \left| \int_{s_n}^{r_n} d\sigma \right| + |r_n - s_n| \right\}$$

$$= 0$$

because of the right-continuity of σ . Thus z is well-defined.

We now show that the convergence is uniform on the rationals.

If possible assume the contrary. Then there is an $\varepsilon_0 > 0$ a rational number t so that for all $N > 0$ there is some $m \geq N$ for which

$$|z_m(t) - z(t)| \geq \varepsilon_0 . \tag{II.1.10}$$

Choose a rational $r > t$ such that

$$|\sigma(t) - \sigma(r)| < \frac{\varepsilon_0}{12a} \tag{II.1.11}$$

where $t < r < t + \delta(\varepsilon_0, t)$. If necessary restrict r further so that

$$t < r < t + \frac{\varepsilon_0}{12} . \tag{II.1.12}$$

Since $z_m(r) \to z(r)$ as $m \to \infty$, there is an $N(\varepsilon, r)$ for which

$$|z_m(r) - z(r)| < \frac{\varepsilon_0}{6} \qquad \text{(II.1.13)}$$

for all $m \geq N_0$.

By our supposition then, for some $m \geq N_0$, such a rational t in K

$$\begin{aligned} \varepsilon_0 &\leq |z_m(t) - z(t)| \\ &\leq |z_m(r) - z(r)| + |z_m(t) - z_m(r)| + |z(r) - z(t)| \\ &\leq |z_m(r) - z(r)| + 2a\left|\int_r^t d\sigma\right| + 2|r - t| \\ &< \frac{\varepsilon_0}{6} + 2a\left(\frac{\varepsilon_0}{12}\right) + 2 \cdot \frac{\varepsilon_0}{12} = \frac{\varepsilon_0}{2} \end{aligned}$$

a contradiction.

In the above we have used the inequality

$$|z(r) - z(t)| < a\left|\int_r^t d\sigma\right| + |r - t|$$

the proof of which is almost immediate since z is a limit of functions each satisfying the same inequality.

Hence the convergence is uniform on the rationals in K.

From this it is now simple to conclude that for given $\varepsilon > 0$, (s_n) as earlier, t fixed,

$$|z_m(s_n) - z(t)| < \varepsilon \qquad m, n \geqq N(\varepsilon) \ . \qquad \text{(II.1.14)}$$

For given $\varepsilon > 0$,

$$|z_m(s_n) - z(t)| < |z_m(s_n) - z(s_n)| + |z(s_n) - z(t)|$$

and the right-continuity of z implies that

$$|z(s_n) - z(t)| < \frac{\varepsilon}{2} \qquad n \geq N_1(\varepsilon)$$

while the uniform convergence of z_m on the rationals implies that

$$|z_m(s_n) - z(s_n)| < \frac{\varepsilon}{2} \qquad m \geq N_2(\varepsilon)$$

where N_2 is independent of n . From these last two inequalities there follows (II.1.14).

Finally, if t is rational,

$$z(t) = \lim_{m\to\infty} z_m(t)$$

by definition while if t is irrational, we let s_n be as before so that

$$z(t) = \lim_{n\to\infty} z(s_n) = \lim_{n\to\infty} \left\{ \lim_{m\to\infty} z_m(s_n) \right\}$$

$$= \lim_{m\to\infty} \left\{ \lim_{n\to\infty} z_m(s_n) \right\}$$

on account of (II.1.14),

$$= \lim_{m\to\infty} z_m(t) \ .$$

Now an argument similar to a previous one shows that the convergence is uniform on the irrationals and so everywhere on K .

Thus for $x_k \in B_n$, $x_k' \equiv z_k$ we have shown the existence of a subsequence which converges uniformly on compact subsets of $[T, T+n)$.

Moreover the uniform convergence of the (x_k') implies that of the (x_k) and thus letting x be their limit,

$$\|x_k - x\|_\infty = \sup_{t\in[T,\infty)} |x_k(t) - x(t)|$$

$$\leqq \max\left\{ \sup_{t\in[T,T+n)} |x_k(t) - x(t)| \ , \quad \sup_{t\in[T+n,\infty)} |x_k(t) - x(t)| \right\} .$$

There is no loss of generality in assuming that T is rational so that $T+n$ is rational and so the $x_k(t)$ can be defined at $t = T+n$ and will converge uniformly to some limit which we use to define $x(T+n)$. We then define $x(t)$ for $t \geq T+n$ by letting $x(t) - x(T+n) = c$.

If we let $x_k(t) = c_k$, $t \geq T + n$, we must then have

$$|c_k - c| \to 0 \qquad k \to \infty .$$

Hence

$$\sup_{t \in [T+n, \infty)} |x_k(t) - x(t)| = |c_k - c| \to 0$$

as $k \to \infty$.

$$\sup_{t \in [T, T+n)} |x_k(t) - x(t)| \to 0$$

as $k \to \infty$ from previous considerations.

Thus $\|x_k - x\|_\infty \to 0$ as $k \to \infty$.

Similarly since $Q(t)$ is bounded away from zero on $[T, T+n)$ and $x_k' \equiv 0$ on $[T+n, \infty)$,

$$\left\| \frac{x_k' - z}{Q} \right\|_\infty = \sup_{t \in [T, \infty)} \left| \frac{x_k'(t) - z(t)}{Q(t)} \right|$$

$$= \sup_{t \in [T, T+n)} \left| \frac{x_k'(t) - z(t)}{Q(t)} \right| \tag{II.1.15}$$

since both x_k' and z are identically zero on $[T+n, \infty)$. Then uniform convergence of the z_k to z implies that (II.1.15) tends to zero as $k \to \infty$. (Also $z = x'$ follows from this.)

Hence

$$\|x_k - x\|_B = \|x_k - x\|_\infty + \left\|\frac{x_k' - x'}{Q}\right\|_\alpha$$

$$\to 0 \qquad k \to \infty ,$$

and so B_n is compact.

APPENDIX III

§1. EIGENVALUES OF GENERALIZED DIFFERENTIAL EQUATIONS

We shall mainly be concerned with some basic results related to eigenvalue problems

$$\ell[y] = \lambda y \qquad \text{(III.1.1)}$$

where $\ell[y]$ was defined in Chapter 3.

THEOREM III.1.0:

The initial value problem

$$\ell[y] = \lambda y \qquad \text{(III.1.2)}$$

$$Y(a) = \alpha \quad , \quad y'(a) = \beta \qquad \text{(III.1.3)}$$

where $\alpha , \beta \in \mathbb{C}$, and λ is a complex parameter has the unique solution $y(x , \lambda)$ and moreover for fixed x , the functions $y(x , \lambda)$, $y'(x , \lambda)$ are entire functions of λ .

Proof: For fixed λ the existence and uniqueness of solutions follows from Theorem 3.2.0 and [3, p. 341]. That $y(x , \lambda)$ is an entire function of λ can be found in [3,

p. 355]. The complete result is found in [35, p. 250 and p. 216, Theorem 2]. We only need to note that the proof in [35] also applies when ν is, more generally, of bounded variation.

THEOREM III.1.1:

Let σ be a non-decreasing right-continuous function such that for some set $E \subset [a, b]$,

$$\int_E d\sigma(t) > 0 . \qquad \text{(III.1.4)}$$

Let ν be a right-continuous function of bounded variation such that for some set $F \subset [a, b]$

$$\int_F |d\nu(t)| > 0 . \qquad \text{(III.1.5)}$$

Then the problem

$$-\frac{d}{d\nu(x)}\left\{y'(x) - \int_a^x y(s)\,d\sigma(s)\right\} = \lambda y(x) \qquad \text{(III.1.6)}$$

$$y(a)\cos\alpha - y'(a)\sin\alpha = 0$$

$$y(b)\cos\beta + y'(b)\sin\beta = 0 \qquad \text{(III.1.7-8)}$$

with $\alpha, \beta \in [0, \pi/2)$ has only real eigenvalues.

Proof: If possible let λ, $\operatorname{Im}\lambda \neq 0$, be an eigenvalue of the problem and $y(x, \lambda)$ the corresponding eigenfunction.

We multiply (III.1.6) by $\bar{y}$ and integrate with respect to ν to obtain

$$\int_a^b \bar{y}(x, \lambda) d\left\{y'(x, \lambda) - \int_a^x y d\sigma\right\} = -\lambda \int_a^b |y(x, \lambda)|^2 d\nu(x) .$$

Integrating the left hand side by parts,

$$[\bar{y}(x, \lambda) y'(x, \lambda)]_a^b = \int_a^b \{|y'(x, \lambda)|^2 dx + |y(x, \lambda)|^2 d\sigma(x)\} - \lambda \int_a^b |y(x, \lambda)|^2 d\nu(x) . \qquad \text{(III.1.9)}$$

Using the boundary conditions (III.1.7-8) we find,

$$-\tan \beta |y'(b, \lambda)|^2 - \tan \alpha |y'(a, \lambda)|^2 = [\bar{y}(x, \lambda) y'(x, \lambda)]_a^b .$$

Combining the latter into the former and taking imaginary parts we obtain

$$(\text{Im } \lambda) \int_a^b |y(x, \lambda)|^2 d\nu(x) = 0 \qquad \text{(III.1.10)}$$

and thus, since $\text{Im } \lambda \neq 0$,

$$\int_a^b |y(x, \lambda)|^2 d\nu(x) = 0 . \qquad \text{(III.1.11)}$$

Inserting the latter into (III.1.9) we must therefore have

$$\int_a^b \{|y'(x\,,\,\lambda)|^2\, dx + |y(x\,,\,\lambda)|^2\, d\sigma(x)\}$$

$$= -\tan\beta\,|y'(b\,,\,\lambda)|^2 - \tan\alpha\,|y'(a\,,\,\lambda)|^2\,. \qquad \text{(III.1.12)}$$

But since $\alpha\,,\,\beta \in [0\,,\,\pi/2)$ the right hand side is strictly negative while the left hand side is necessarily non-negative. Hence we must have

$$\int_a^b \{|y'(x\,,\,\lambda)|^2\, dx + |y(x\,,\,\lambda)|^2\, d\sigma(x)\} = 0\,.$$

From the latter it follows that $y(x\,,\,\lambda)$ is a constant on $[a\,,\,b]$ and (III.1.4) finally implies that this constant must be zero. Hence $y(x\,,\,\lambda) \equiv 0$ on $[a\,,\,b]$ which is contrary to the requirement that y is an eigenfunction. This contradiction therefore implies $\text{Im}\,\lambda = 0$, hence all eigenvalues are real.

<u>THEOREM III.1.2:</u>

Let σ be a right-continuous function of bounded variation and suppose that $\nu(x)$ is right-continuous and non-decreasing and, additionally, that it satisfy (III.1.5) on some set $F \subset [a\,,\,b]$. Then the problem (III.1.6-7-8) has only real eigenvalues if $\alpha\,,\,\beta \in [0\,,\,\pi)$.

<u>Proof:</u> We proceed as in the proof of the previous theorem by assuming the existence of a non-real eigenvalue λ and

arguing, as in that proof, until we reach (III.1.9). If $\alpha, \beta \neq \pi/2$, we take imaginary parts in (III.1.9) obtaining (III.1.10). Since $\operatorname{Im} \lambda \neq 0$,

$$\int_a^b |y(x, \lambda)|^2 \, d\nu(x) = 0 .$$

But since $y \neq 0$ and ν satisfies (III.1.5) this is impossible unless

$$\int_F |y(x, \lambda)|^2 \, d\nu(x) = 0 .$$

Since F has positive ν-measure $y(x, \lambda) \equiv 0$ on F. Since y is a solution of (III.1.6) then y must be the trivial solution. This contradiction shows that no non-real eigenvalues can exist. The case when either or both of $\alpha, \beta = \pi/2$ is treated separately with a similar argument to the one above.

We note here that if ν is an arbitrary function of bounded variation on $[a, b]$ then the eigenvalues of (III.1.6-7-8) need not be all real even if $\alpha, \beta \in [0, \pi/2]$. The latter situation was illustrated in Chapter 4. For the singular case see the papers [10], [11].

§2. LINEAR OPERATORS IN A HILBERT SPACE:

For the basic notions regarding Hilbert spaces one may refer to any book on functional analysis. The space

$L^2(V; I)$ defined by those (equivalence classes of) functions such that

$$\|f\| \equiv \left\{ \int_I |f(x)|^2 \, dV(x) \right\}^{\frac{1}{2}} < \infty \qquad \text{(III.2.0)}$$

is a Hilbert space whenever $V(x)$ is a non-decreasing function.

If H_1, H_2 are two Hilbert spaces their *direct sum*, $H_1 \oplus H_2$ consists of all pairs $\{x_1, x_2\}$ where $x_1 \in H_1$, $x_2 \in H_2$.

A linear operator A in a Hilbert space H is said to be *closed* if given any sequence $x_n \in D_A$ (domain of A) such that $x_n \to x$, $Ax_n \to y$ then $x \in D_A$ and $y = Ax$.

An operator A *admits a closure* $\tilde{A}$ if and only if the relations

$$x_n \in D_A \,,\; x_n' \in D_A \,,\; x_n \to x \,,\; x_n' \to x \,,\; Ax_n \to y \,,\; Ax_n' \to y'$$

imply that $y = y'$.

The domain of the closure $D(\tilde{A})$ consists precisely of those vectors x for which there is $x_n \in D_A$ satisfying

$$x_n \to x \qquad \{Ax_n\} \text{ converges as } n \to \infty .$$

Then for $x \in D(\tilde{A})$, $\tilde{A}x \equiv \lim_{n\to\infty} Ax_n$.

A set $S \subset H$ is *dense* if its closure $\tilde{S} = H$. In fact a subspace S is dense in H if and only if there is no nonzero vector in H which is orthogonal to S .

Let A be any linear operator with a dense domain of definition in H . Denote the domain of A by D .

The set of all vectors y such that

$$(Ax , y) = (x , z)$$

holds for all $x \in D$ and some $z \in H$, defines the domain D^* of the operator *adjoint* to A , denoted by A^* , and defined by

$$y \in D^* \qquad A^*y = z .$$

A^* is simply called the adjoint of A .

An operator A is said to be *hermitian* if, for all $x , y \in D$,

$$(Ax , y) = (x , Ay) .$$

A hermitian operator with a dense domain of definition is said to be *symmetric*.

An operator with a dense domain of definition is said to be *self-adjoint* if $A = A^*$.

A complex number λ is a *regular point* of the operator A if the inverse $(A - \lambda I)^{-1}$ exists and represents a bounded operator defined on the whole space H . All non-regular points λ are called *points of the spectrum* of A .

The set of all eigenvalues of A constitutes the *discrete spectrum* of A.

All other points of the spectrum (if any) are called *points of the continuous spectrum* (essential spectrum) of A. The collection of all such points constitutes the *continuous spectrum* of A. The *spectrum* of A is the union of the discrete and continuous spectra of A.

The spectrum of any self-adjoint operator A in a Hilbert space is real.

For the general theory of extensions of symmetric operators we refer to [46, §14].

§3. LINEAR OPERATORS IN A KREIN SPACE:

For general information concerning indefinite inner product spaces and, in particular, Krein spaces we refer to [7] and [40].

A linear space K with a generally indefinite inner product $[\ ,\ ]$ is called a *Krein space* if

$$K = K_+ [\dot{+}] K_-$$

where $K_+ (K_-)$ is a Hilbert space with respect to the inner product $[\ ,\ ]$ $(-[\ ,\ ])$ respectively. The symbol $[\dot{+}]$ denotes a direct sum which is orthogonal with respect to the inner product $[\ ,\ ]$, i.e.

$$K_+ \cap K_- = \{0\} \quad , \quad [f_+ , f_-] = 0$$

whenever $f_+ \in K_+$, $f_- \in K_-$.

A positive definite inner product $(\, , \,)$ can be defined on K by

$$(f , g) = [f_+ , g_+] - [f_- , g_-]$$

where $f = f_+ + f_-$, $g = g_+ + g_-$, $f_\pm , g_\pm \in K_\pm$.

$(K , (\, , \,))$ is then a Hilbert space and if $P_\pm$ denotes the orthogonal projector of K on $K_\pm$ then

$$[f , g] = (Jf , g)$$

where $J = P_+ - P_-$ and $P_+ + P_- = I$. The important fact here is that different decompositions of K generate different inner products $(\, , \,)$ but the corresponding norms are all equivalent [7]. Topological concepts in a Krein space are to be interpreted in the norm topology of the induced Hilbert space.

<u>Example</u>: Let $\nu(x)$ be a right-continuous function locally of bounded variation on $[a , \infty)$, $a > -\infty$. Let m_ν $(|m_\nu| \equiv V)$ denote the signed measure (measure) induced by the function ν respectively (Chapter 3, section 1). V is called the total variation measure of ν and when L^2 is equipped with such a measure and the norm is defined by

$$\|f\| \equiv \int_a^\infty |f(x)|^2 \, dV(x) \quad < \infty$$

then $L^2(V\,;\,(a\,,\,\infty)) \equiv K$ is a Hilbert space with inner product

$$(f\,,\,g) \equiv \int_a^\infty f(x)\bar{g}(x)\,dV(x)$$

for $f\,,\,g \in K$. If one introduces the indefinite inner product

$$[f\,,\,g] \equiv \int_a^\infty f(x)\bar{g}(x)\,d\nu(x)$$

for $f\,,\,g \in K$, then K becomes a Krein space. For since the measure m_ν is absolutely continuous with respect to $|m_\nu|$ the Radon-Nikodym derivative $dm_\nu/d|m_\nu|$ exists $|m_\nu|$-almost everywhere and is equal to (Chapter 3.1)

$$\frac{d\nu(x)}{dV(x)} \equiv \lim_{h\downarrow 0} \left\{\frac{\nu(x+h) - \nu(x-h)}{V(x+h) - V(x-h)}\right\}$$

$|m_\nu|$-almost everywhere and moreover

$$\int_a^\infty f(x)\bar{g}(x)\,d\nu(x) = \int_a^\infty f(x)\bar{g}(x)\,\frac{d\nu(x)}{dV(x)}\,dV(x) \ .$$

(See for example [24, p. 134, Theorem B and (1), p. 135].)

Thus, in this case, for $f \in K$

$$(Jf)(x) = \frac{d\nu(x)}{dV(x)} \cdot f(x) \qquad [V] \ ,$$

and

$$(P_{\pm}f)(x) = \frac{1}{2}\left\{1 \pm \frac{d\nu(x)}{dV(x)}\right\} f(x) \qquad [V]$$

where [V] means $|m_\nu|$-almost everywhere.

It is then readily seen that

$$[f , g] = (Jf , g)$$

with J defined above and $K_{\pm} \equiv P_{\pm}K$ by definition. If A is an operator with dense domain of definition in the Krein space K, the *J-adjoint* A^x is characterized by the relation

$$[Af , g] = [f , A^x g] \qquad f \in D(A) , g \in D(A^x) .$$

In this sense it is defined analogously as in the Hilbert space setting, for $D(A^x)$ still consists of those $g \in K$ such that

$$[Af , g] = [f , h]$$

holds for all $f \in D(A)$ and some $h \in K$. Then $A^+g \equiv h$.

The closed operator A is called *J-symmetric* if it is symmetric with respect to [,] , i.e.

$$[Af , g] = [f , Ag] \qquad f , g \in D(A)$$

and $D(A)$ is dense in K.

The operator A with $D(A)$ dense in K is *J-self-adjoint* if $A = A^x$.

In contrast with self-adjoint operators in Hilbert spaces, J-self-adjoint operators may have non-real spectrum (see for example, [7, p. 133, Example 6.4]).

If A is a densely defined operator in a Krein space and A^x is its Krein space adjoint (J-adjoint) then its Hilbert space adjoint A^* is defined as usual replacing $[\ ,\ ]$ by $(\ ,\ )$ in the defining relation for the adjoint. The adjoints A^x , A^* are related by the formula

$$A^x = JA^*J$$

where $J = J^*$ and $J^2 = I$. For this result see [45, p. 122]. Moreover if S , T are operators in the Krein space with adjoints S^x , T^x then [45, p. 122],

$$(ST)^x \supset T^x S^x \ .$$

§4. FORMALLY SELF-ADJOINT EVEN ORDER DIFFERENTIAL EQUATIONS WITH AN INDEFINITE WEIGHT-FUNCTION:

In the following we shall assume that

$$p_k \in C^{(n-k)}(a\ ,\ b) \qquad k = 0\ ,\ \dots\ ,\ n$$

and $p_0(x) > 0$ on $[a\ ,\ b]$.

The boundary problem

$$\sum_{k=0}^{n} (-1)^{n-k}\left(p_k f^{(n-k)}\right)^{(n-k)} = \lambda f \qquad \text{(III.4.1)}$$

$$f^{(j)}(a) = f^{(j)}(b) = 0 \qquad j = 0, \ldots, n-1 \qquad \text{(III.4.2)}$$

is self-adjoint [9, p. 201, Ex. 3]. Consequently the eigenvalues are real, bounded below, and have no finite point of accumulation. Also eigenfunctions corresponding to distinct eigenvalues are orthogonal in $L^2(a, b)$. Moreover, if $f \in C^n(a, b)$ satisfies (III.4.2) then f can be expanded into a uniformly convergent series of the eigenfunctions $y_k(x)$ of (III.4.1-2), i.e.,

$$f(x) = \sum_{k=0}^{\infty} (f, y_k) y_k(x) \qquad \text{(III.4.2)}$$

where we assume that $\|y_k\| = 1$ for all k. (For the latter result, see [9, p. 197, Theorem 4.1].)

As usual we define an operator A with domain

$$D(A) = \{f \in C^n(a, b)\}$$

such that

$$Af = \sum_{k=0}^{n} (-1)^{n-k}\left(p_k f^{(n-k)}\right)^{(n-k)} . \qquad \text{(III.4.3)}$$

If we let

$$D(\tilde{A}) = \{f \in D(A) : f^{(i)}(a) = f^{(i)}(b) = 0 \quad \text{for} \quad i = 0, 1, \ldots, n-1\}$$

and

$$\tilde{A}f = Af \qquad f \in D(\tilde{A})$$

then integrating $(\tilde{A}f , f)$ by parts we find

$$(\tilde{A}f , f) = \int_a^b \sum_{j=0}^{n} p_{n-j} |f^{(j)}|^2 \, dx \qquad f \in D(\tilde{A}) . \tag{III.4.4}$$

Now from (III.4.2) and the completeness of the eigenfunctions $y_k(x)$ in $L^2(a , b)$ we also have, for $f \in D(\tilde{A})$

$$(\tilde{A}f , f) = \sum_{k=0}^{\infty} \lambda_k |(f , y_k)|^2 . \tag{III.4.5}$$

Thus if we let $\lambda_0 , \lambda_1 , \dots , \lambda_{N-1}$ be the N distinct negative eigenvalues of (III.4.1-2) and if for some f we have $(f , y_k) = 0$, $k = 0 , 1 , \dots , N-1$, then

$$(\tilde{A}f , f) > 0 .$$

The rest of the argument now follows that in Chapter 4, §2. For, an adaptation of Lemma 4.2.2 shows that, if λ , μ are non-real eigenvalues of

$$Af = \lambda rf \qquad f^{(i)}(a) = f^{(i)}(b) = 0 \qquad i = 0 , \dots , n-1 , \tag{III.4.6}$$

where $r(x)$ is, say, continuous on $[a , b]$ and changes sign at least once there, and $\lambda \neq \bar{\mu}$, then

$$\int_a^b f(x)\bar{g}(x)r(x)dx = 0 \qquad \text{(III.4.7)}$$

$$\int_a^b \sum_{j=0}^{n} P_{n-j} f^{(j)}\bar{g}^{(j)} dx = 0 \qquad \text{(III.4.8)}$$

where f, g are the eigenfunctions corresponding to λ, μ respectively.

Thus we let $\mu_0, \ldots, \mu_{M-1}$ be the non-real eigenvalues of (III.4.6) such that $\mu_i \neq \bar{\mu}_j$, $0 \leqq i, j \leqq M-1$, with eigenfunctions $\phi_0, \phi_1, \ldots, \phi_{M-1}$. Since $\phi_i(x) \in D(\tilde{A})$ we have

$$\int_a^b \sum_{j=0}^{n} P_{n-j} |\phi_i^{(j)}|^2 dx = \sum_0^\infty \lambda_k |(\phi_i, y_k)|^2$$

for $i = 0, \ldots, M-1$. Thus we let

$$f(x) = \sum_{j=0}^{M-1} e_j \phi_j(x) \qquad \text{(III.4.9)}$$

and, as in Chapter 4.2, we see that if $M > N$ then it is possible to choose the coefficients e_j such that $(f, y_k) = 0$, $k = 0, \ldots, N-1$. This would then imply that

$$(\tilde{A}f, f) > 0 .$$

But by substituting (III.4.9) in the latter relation and

expanding the form, we shall find that

$$(\tilde{A}f , f) = 0$$

on account of (III.4.7-8). This contradiction then proves the result.

*Note: The problem here is the following: Richardson's idea is to approximate the eigenvalues of the continuous problem,

$$(py')' + (q + \lambda k)y = 0$$

$$y(0) = y(1) = 0$$

by the eigenvalues of the discrete problem,

$$m^2 \Delta(p_i \Delta y_{i-1}) + q_i y_i + \lambda k_i y_i = 0 ,$$

$i = 0, 1, 2, \ldots, m$ where the values of y, p, q, k at the points i/m are denoted by y_i, p_i, q_i, k_i respectively. The claim appears to be that for large values of m the eigenvalues of the discrete problem with

$$y_0 = y_m = 0$$

are approximations to the eigenvalues of the above continuous problem. However it is not at all clear that if the discrete problem has non-real eigenvalues then these must necessarily approximate non-real eigenvalues in the continuous case. For it is conceivable that these limits may be real. It does not seem as if enough information is provided in [53] to exclude the latter possibility. In fact in some cases no non-real eigenvalues may exist and so one needs to establish some criteria on the coefficients which will guarantee their existence.

BIBLIOGRAPHY

[1] APOSTOL, T., *Mathematical Analysis*, Second Edition, Addison-Wesley, Massachusets, 1974.

[2] ATKINSON, F.V., *On second order nonlinear oscillation*, Pacific J. Math. 5, (1955), 643-647.

[3] ________, *Discrete and Continuous Boundary Problems*, New York, Academic Press, 1964.

[4] LANGER, H., *Sturm-Liouville Problems with Indefinite Weight Function and Operators in Spaces with an Indefinite Metric*, Uppsala Conference on Differential Equations 1977, Almqvist & Wiksell, pp. 114-124.

[5] BÔCHER, M., *Boundary problems and Green's functions for linear differential and linear difference equations*, Ann. Math. (2), 13, (1911-12), 71-88.

[6] ________, *Boundary problems in one dimension*, 5th International Congress of Mathematicians, Proceedings, Cambridge U.P., 1912, Vol. 1, 163-195.

[7] BOGNÁR, J., *Indefinite Inner Product Spaces*, Berlin, Springer, 1974.

[8] BUTLER, G.J., *On the oscillatory behavior of a second order nonlinear differential equation*, Ann. Mat. Pura ed Appl. Seria IV, Tomo CV, (1975), 73-91.

[9] CODDINGTON, E.A. and N. Levinson, *Theory of Ordinary Differential Equations*, McGraw-Hill, New York, 1955.

[10] DAHO, K. and H. Langer, *Some remarks on a paper by W.N. Everitt*, Proc. Royal Soc. Edinb. 78A, (1977), 71-79.

[11] ________, *Sturm-Liouville operators with an indefinite weight-function*, Proc. Royal Soc. Edinb. 78A, (1977), 161-191.

[12] DERR, V. Ya, *Criterion for a difference equation to be non-oscillatory*, (Russian), Diff. Urav, 12, (4), (1976), 747-750 or Differential Equations, 12, (4), April 1976, 524-527.

[13] DUNKEL, O., *Some applications of Green's function in one dimension*, Am. Math. Soc. Bul., 2nd series, 8, (1901-02), 288-292.

[14] EVERITT, W.N., *Self-adjoint boundary value problems on finite intervals*, J. Lond. Math. 37, (1962), 372-384.

[15] ________, *A note on the self-adjoint domains of second order differential equations*, Q.J. Math., Oxford (2), 14, (1963), 41-45.

[16] ________, *On the spectrum of a second order linear differential equation with a p-integrable coefficient*, App. Anal., 2, 1972, 143-160.

[17] ________, *A note on the Dirichlet condition for second order differential expressions*, Can. J. Math. 28, (2), (1976), 312-320.

[18] EVERITT, W.N., M. Giertz, J. Weidmann, *Some remarks on a separation and limit-point criterion of second-order ordinary differential expressions*, Math. Annal. 200, (1973), 335-346.

[19] FITE, W., *Concerning the zeros of the solutions of certain differential equations*, Trans. Am. Math. Soc. 19, (1918), 341-352.

[20] FORT, T., *Oscillatory and non-oscillatory linear difference equations of the second order*, Quart. J. Pure Appl. Math. 45, (1914-15), 239-257.

[21] ________, *Finite Differences and Difference Equations in the Real Domain*, Oxford U.P., 1948.

[22] GANELIUS, T., *Un théorème taubérien pour la transformée de Laplace*, Acad. d. Sci. Paris Compt. Rend., 242, (1956), 719-721.

[23] GLAZMAN, I.M., *Direct Methods of Qualitative Spectral Analysis of Singular Differential Operators*, Israel Program for Scientific Translations (IPST), 1965.

[24] HALMOS, P.R., *Measure Theory*, Van Nostrand Reinhold, New York, 1950.

[25] HARTMAN, P., *Ordinary Differential Equations*, New York, Wiley, 1964.

[26] HARTMAN, P. and A. Wintner, *Oscillatory and non-oscillatory linear differential equations*, Am. J. Math. 71, (1949), 627-649.

[27] ________, *On non-oscillatory linear differential equations with monotone coefficients*, Am. J. Math. 76, (1954), 207-219.

[28] HELLWIG, G., *Differential Operators of Mathematical Physics*, Springer, Berlin, New York, 1964.

[29] HILBERT, D., *Grundzüge einer allgemeinen Theorie der linearen Integralgleichungen*, Göttinger Nachrichten, 1 und 2 Mitteilung (1904), 4 and 5, (1906).

[30] HILDEBRAND, F.B., *Finite Difference Equations and Simulations*, Prentice-Hall, New Jersey, 1968.

[31] HILLE, E., *Non-oscillation theorems*, Trans. Am. Math. Soc., 64, (1948), 234-252.

[32] HINTON, D., and R.T. Lewis, *Spectral analysis of second order difference equations*, J. Math. Anal. Appl. 63, 2, (1978), 421-438.

[33] INCE, E.L., *Ordinary differential equations*, Dover, New York, 1956.

[34] JÖRGENS, K., *Spectral Theory of Second-Order Ordinary Differential Equations*, Matematisk Institut, Aarhus Universitet, 1964.

[35] KAC, I.S., *The existence of spectral functions of generalized second-order differential systems with a boundary condition at a singular end*, (Russian), Mat. Sb. 68, (110), (1965), 174-227, or Translations Am. Math. Soc., series 2, Vol. 62, (1967), 204-262.

[36] ________, *A remark on the article "The existence of spectral functions of generalized second order differential systems with a boundary condition at a singular end"*, Mat. Sb. 76, (118), (1968), No. 1, or Math. USSR. Sbornik, Vol. 5, (1968), No. 1, 141-145.

[37] ________, *On the completeness of the system of eigenfunctions of generalized linear differential expressions of the second order*, (Russian), Dokl. Akad. Nauk. Armyan. LX, 4, (1975), 198-203.

[38] KAC, I.S., and M.G. Krein, *On the spectral functions of the string*, Translations Am. Math. Soc., (2), Vol. 103, 1974.

[39] KREIN, M.G., *On a generalization of investigations of Stieltjes*, Doklady Akad. Nauk. SSSR 87, (1952), 881-884 (Russian).

[40] KREIN, M.G., *Introduction to the geometry of indefinite J-spaces and to the theory of operators in those spaces*, Second Math. Summer School, Part I, Naukova Dumka, Kiev, 1965, 15-92 or Translations Am. Math. Soc. (2), 93, (1970), 103-176.

[41] LANGER, H., *Zur Spektraltheorie verallgemeinerter gewönlicher differentialoperatoren zweiter ordnung mit einer nichtmonotonen gewichtsfunktion*, University of Jyväskylä, Dept. of Math., Report 14, (1972).

[42] LEIGHTON, W., *Comparison theorems for linear differential equations of second order*, Am. Math. Soc. Proc., 13, (1962), 603-610.

[43] MASON, M., *On boundary value problems of linear ordinary differential equations of second order*, Trans. Am. Math. Soc., 7, (1906), 337-360.

[44] MOORE, R.A., *The behavior of solutions of a linear differential equation of the second order*, Pacific J. Math., 5, No. 1, (1955), 125-145.

[45] MOULTON, E.J., *A theorem in difference equations on the alternation of nodes of linearly independent solutions*, Ann. Math., (2), 13, (1912), 137-139.

[46] NAIMARK, M.A., *Linear Differential Operators*, Vol. 2, Ungar, New York, 1968.

[47] OPIAL, Z., *Sur les intégrales oscillantes de l'équation différentielle* $u'' + f(t)u = 0$, Ann. Pol. Mat., 4, (1957-58), 308-313.

[48] PICONE, M., *Sui valori eccezionali di un parametro da cui dipende un'equazione differenziale lineare ordinaria del second'ordine*, Annali Scuola N. Sup., Pisa, Scienze Fisiche e Matematiche, Seria 1, 11, (1909), 1-141.

[49] PORTER, M.B., *On the roots of functions connected by a linear recurrent relation of the second order*, Ann. Math., (2), Vol. 3, (1901-02), 55-70.

[50] REID, W.T., *A criterion of oscillation for generalized differential equations*, Rocky Mtn. J. Math., 7, (1977), 799-806.

[51] RICHARDSON, R.G.D., *Das Jacobische Kriterium der Variationsrechnung und die Oszillationseigenschaften linearer Differentialgleichungen 2. ordnung*, Math. Ann., 68, (1910), 279-304.

[52] ________, *Theorems of oscillation for two linear differential equations of the second order with two parameters*, Trans. Am. Math. Soc., 13, (1912), 22-34.

[53] ________, *Contribution to the study of oscillation properties of linear differential equations of the second order*, Am. J. Math., 40, (1918), 283-316.

[54] RIESZ, M., *Sur les ensembles compacts de fonctions sommables*, Acta Sci. Math. Szeged., 6, (1933), 136-142.

[55] ROYDEN, H.L., *Real Analysis*, Second Edition, Macmillan, New York, 1968.

[56] SANLIEVICI, S.M., *Sur les équations différentielles des cordes et des membranes vibrantes*, Annales de l'Ecole Norm. Sup. Paris, t25, (1909), 19-91.

[57] SMART, D.R., *Fixed Point Theorems*, Cambridge U.P., London, New York, , CT, No. 66.

[58] STURM, C., Mémoire: *Sur les équations différentielles linéaires du second ordre*, J. Math. Pures Appl., (1), 1, (1836), 106-186.

[59] SWANSON, C.A., *Comparison and Oscillation Theory of Linear Differential Equations*, Academic Press, New York, 1968.

[60] TAAM, C-T., *Non-oscillatory differential equations*, Duke Math. J., 19, (1952), 493-497.

[61] WINTNER, A., *A criterion of oscillatory stability*, Quart. Appl. Math., 7, (1949), 115-117.

[62] ________, *On the non-existence of conjugate points*, Am. J. Math., 73, (1951), 368-380.

[63] ________, *On the comparison theorem of Kneser-Hille*, Math. Scand., 5, (1957), 255-260.

[64] F.V. Atkinson, W.N. Everitt, K.S. Ong, *On the m-coefficient of Weyl for a differential equation with an indefinite weight-function*, Proc. London Math. Soc., (3), 29, (1974), 368-384.

[65] W.N. Everitt, *Some remarks on a differential expression with an indefinite weight-function*, in Spectral Theory and asymptotics of differential equations. Math. Stud., 13 (Amsterdam: North-Holland, 1974).

[66] W.N. Everitt and C. Bennewitz, *Some remarks on the Titchmarsh-Weigh m-coefficient*, in A Tribute to Åke Pleijel, Uppsala Universitet, (1980), 49-108.

[67] W.N. Everitt, M.K. Kwong, A. Zettl, *Oscillation of eigenfunctions of weighted regular Sturm-Liouville problems*, To appear in J. London Math. Soc.

[68] W. Feller, *Diffusion processes in one dimension*, Trans. Amer. Math. Soc., 77, (1954), 1-31.

[69] _________, *On second order differential operators*, Ann. of Math., (2), 61, (1955), 90-105.

[70] _________, *On differential operators and boundary conditions*, Comm. Pure Appl. Math., (8), (1955), 203-216.

[71] _________, *On generalized Sturm-Liouville operators*, in Proceedings of the conference on differential equations, (dedicated to A. Weinstein), University of Maryland Book Store, College Park, Md., (1956), 251-270.

[72] _________, *Generalized second order differential operators and their lateral conditions*, Illinois J. Math., 1, (1957), 459-504.

[73] _________, *On the intrinsic form for second order differential operators*, Illinois J. Math., 2, (1958), 1-18.

[74] Mingarelli A.B., *Some extensions of the Sturm-Picone theorem*, C.R. Math. Rep. Acad. Sci. Canada., 1, (4), (1979), 223-226.

[75] ____________, *A limit-point criterion for a three-term recurrence relation*, C.R. Math. Rep. Acad. Sci. Canada, 3, (1981), 171-175.

[76] ____________, *Indefinite Sturm-Liouville problems*, To appear in the Proceedings of the Conference on Differential Equations 1982, University of Dundee, Scotland, Lecture Notes in Mathematics, Springer-Verlag, N.Y.

[77] ____________, *Some remarks on the order of an entire function associated with a second order differential equation*, To appear in 'Tribute to F.V. Atkinson', Proceedings of the Symposium on Differential Operators, University of Dundee, Scotland, Lecture Notes in Mathematics, Springer-Verlag, N.Y.

[78] ____________, *Asymptotic distribution of the eigenvalues of non-definite Sturm-Liouville problems*, To appear in 'Tribute to F.V. Atkinson', Proceedings of the Symposium on Differential Operators, University of Dundee, Scotland, Lecture Notes in Mathematics, Springer-Verlag N.Y.

[79] W.T. Reid, *Generalized linear differential systems*, Journal of Math. and Mech., 8, (1959) 705-726.

[80] ____________, *Generalized linear differential systems and related Riccati matrix integral equations*, Illinois J. Math., 10, (1966), 701-722

[81] H. Weyl, *Über gewöhnlicher Differentialgleichungen mit singularitäten und die zugehörigen Entwicklungen willkürlioher Functionen*, Mat. Ann., 68, (1910), 220-269.

Subject Index